Reconstructed Pliocene landscape of the Lothagam area in Kenya. In the foreground is *Stegotetrabelodon orbus*, one of the earliest members of the Elephantidae. Behind is *Primelephas gomphotheroides*, a primitive elephantine proboscidean.

TRANSACTIONS

OF THE

AMERICAN PHILOSOPHICAL SOCIETY

HELD AT PHILADELPHIA
FOR PROMOTING USEFUL KNOWLEDGE

NEW SERIES—VOLUME 63, PART 3
1973

ORIGIN AND EVOLUTION OF THE ELEPHANTIDAE

VINCENT J. MAGLIO

Department of Geological and Geophysical Sciences, Princeton University

THE AMERICAN PHILOSOPHICAL SOCIETY
INDEPENDENCE SQUARE
PHILADELPHIA

June, 1973

"15. Now look at Behemoth, your contemporary in my creation.
   He eats grass like an ox.

16. Lo, now his strength is in his loins,
   And his force is in the muscles of his belly.

17. His tail is like a cedar.
   The sinews of his thighs are knit together.

18. His bones are like brass cylinders;
   His ribs are like bars of iron.

19. He is the masterpiece of My animal creation:
   The One that made him furnished him alone with a sword.

20. Surely the mountains bring forth food to him,
   where all the beasts of the field do play.

21. He lies under the lotus tree,
   In the covert of the reed and fen.

22. The lotus trees cover him with their shade;
   The willows of the brook compass him about.

23. Behold if a river overflow, he trembles not:
   He is confident, though Jordan swell even to his mouth.

24. Shall any capture him when he is alert?
   Or snare him by his nose?"

Job 40: 15–24 (Paraphrase)

Copyright © 1973 by The American Philosophical Society

Library of Congress Catalog Card Number 73–75474
International Standard Book Number 0–87169–633–9
US ISSN 0065–9746

PREFACE

During the summer of 1965 an expedition of the Museum of Comparative Zoology, under the direction of Professor Bryan Patterson, discovered a new locality of Pliocene [1] age exposed along the Kanapoi River in the western Rift Valley, Turkana, Kenya. Two years later a similar expedition worked an even earlier deposit at Lothagam Hill, 40 miles north of Kanapoi. Although these two deposits were clearly seen to be important, the great significance of their faunas was not fully realized, even by those of us who worked them, until detailed studies began. The reason was simply that these were the first good fossiliferous deposits of Pliocene age known in Africa. There were no other collections against which they could be compared. Although several excellent Pleistocene deposits were then known, the resolution of faunal and radiometric correlation was too poor for the earlier ages of Kanapoi and Lothagam to be immediately apparent.

The knowledge gained from these localities has allowed a meaningful re-evaluation of other, less complete faunas. The result is that a well-documented biostratigraphic sequence in Africa can now be extended back to about 6.0 million years. A number of other deposits discovered in Africa during the past several years have added further to the known record, extending our knowledge into the later Miocene. Sophisticated radiometric determinations are now available as control points for a time calibration of Pliocene and Pleistocene deposits both in Africa and in Europe.

This rapid increase in our knowledge of the Pliocene epoch in Africa and its relation to the European and Asiatic sequences has made available new opportunities for investigation into the earliest history of the modern faunas of that continent. The Pliocene epoch was a crucial period of time for the origin and early radiation of many modern groups of mammals such as the elephants, hippopotami, the recent groups of suids, many of the African bovids, the African rhinoceroses and the modern giraffes. A study of the emergence of these new adaptive types and their initial radiation was not previously possible. The present opportunity will allow a greater understanding of the recent African fauna. This is one of the last major faunal assemblages intact on the earth, and any study that can add to our knowledge of it and of the ecological relationships between its various components will be of immense value in its ultimate preservation. For the complete understanding of any group of animals, its evolutionary history, paleoecology, and zoogeography are equal in importance to its present adaptations, population structure and distribution.

Because the abundant fossil Proboscidea in the newly discovered Pliocene of Africa appeared to answer many important questions relating to the origin of the Elephantidae and to the evolution of their peculiar adaptations, and because of the recent advances in geochronology it was decided to attempt a re-evaluation of this group in Africa. It was soon clear, however, that any attempt to study the African elephantids without a consideration of Eurasiatic forms would be doomed to the same type of provincial thinking that has plagued the study of so many groups of mammals.

The present investigation represents an effort at bringing together the latest ideas on the correlation of Pliocene/Pleistocene deposits and the origin, early evolution, and zoogeography of the fossil elephants of the world. Many of the concepts presented here are the direct result of the research of many recent workers on all aspects of Pliocene/Pleistocene prehistory. Many other ideas are new and not always in accordance with the philosophy or taxonomic usage of most workers. The single most important criterion employed here is the usefulness and stability of the conceptual framework against which the family can be studied. As new information becomes available changes will doubtless be needed. Nevertheless, I feel that the evidence at this time requires an interpretation close to that presented here.

If this study accomplishes nothing else, it will at least point out the great amount of work still needed on this group. This is especially true for population studies on the many large collections from single localities. Every aspect of elephant evolution touched upon here requires much more intensive analysis.

It is hoped that the sins of the present study will create sufficient stimulus for continued work on this group by other scientists.

---

[1] Recent radiometric calibration of marine sequences containing planktonic foraminifera and molluscan faunas place the Miocene/Pliocene boundary close to the Zone N.18/Zone N.19 level (Van Couvering, 1972), with an estimated age of 5.5 to 5.0 million years (Ikebe, 1969; Berggren, 1969). Although the Pliocene/Pleistocene boundary remains in dispute, the base of the Calabrian in Italy should probably be retained for the present. This is approximately equal to the base of foraminiferal Zone N. 22 (Blow, 1969) and correlates with late Villafranchian European mammalian faunas dated at about 1.9 m.y. (Savage and Curtis, 1970).

Thus, for purposes of this study, the base of the Pliocene epoch is taken at 5.5 m.y. and the top at 2.0 m.y.

ORIGIN AND EVOLUTION OF THE ELEPHANTIDAE

VINCENT J. MAGLIO

CONTENTS

	PAGE
I. Introduction	5
The family Elephantidae	5
History	6
Scope and purpose	7
Abbreviations	8
II. Morphological criteria	8
Nature of the morphological characters	8
Metrical procedure	11
III. The species and genus concepts	13
IV. Systematic revision	14
Elephantidae	16
Stegotetrabelodontinae	16
Stegotetrabelodon	16
S. syrticus	17
S. orbus	18
S. sp	20
Elephantinae	20
Primelephas	20
P. gomphotheroides	21
P. korotorensis	22
Loxodonta	22
L. adaurora	23
L. atlantica	25
L. africana	29
Elephas	31
E. ekorensis	33
E. recki	34
E. iolensis	37
E. namadicus	40
E. falconeri	42
E. planifrons	42
E. celebensis	46
E. platycephalus	47
E. hysudricus	48
E. hysudrindicus	49
E. maximus	50
Mammuthus	50
M. subplanifrons	51
M. africanavus	53
M. meridionalis	53
M. armeniacus	57
M. primigenius	60
North American Mammoths	61
M. meridionalis	63
M. imperator	63
M. columbi	63
M. primigenius	63
Elephantinae, gen. indet	63
Identification of elephant species	64
V. Correlation of Pliocene/Pleistocene deposits	67
VI. Origin and phylogeny	76
VII. Evolutionary trends	87
Evolution and function of the elephant dentition	87
Evolution of the elephant mandible	95
Evolution of the elephant skull	97
Function and evolution of the masticatory apparatus	101
VIII. Rates of evolution in the Elephantidae	105
IX. Zoogeography	111
X. Summary	118
XI. Acknowledgments	119
XII. Literature cited	121
XIII. Index	145

I. INTRODUCTION

THE FAMILY ELEPHANTIDAE

Diversity within the Proboscidea has resulted in a wide range of forms and adaptations to a variety of ecological situations. The history of the order was characterized by a series of adaptive shifts that produced such varied groups as the mastodonts, gomphotheres, stegodonts, and elephants.[2] The lack of agreement as to the content of these groups has led to a misunderstanding of the distinctive adaptations represented. In the present work the term mastodont includes only the family Mammutidae and is not used as a common term for the Gomphotheriidae as it almost always has. For the latter, the term gomphothere is applied below. The stegodonts were a specialized group probably derived from the mastodonts and evolving in parallel with the true elephants, although they were unrelated (Maglio and Hendey, 1970). In the following pages the term elephant will be restricted to the family Elephantidae. Within this family the only common name used is that of the mammoth which is applied to all members of the genus *Mammuthus*. Figure 1 presents a generalized phyletic scheme showing the relationships between the various groups of Proboscidea.

The greatest diversity within the order is found among the Gomphotheriidae. Strikingly specialized groups include the amebelodonts with "shovel-tusked" lower jaws and the beak-tusked rhynchotheres. Yet, despite these differences (probably related to food gathering), the family is remarkably uniform in the structure of its dentition. Except for the short-jawed gomphotheres, such as in the subfamily Anancinae in which some cranial specializations paralleled those of the true elephants, the mechanics of mastication seem to have been fairly constant within the family.

The emergence of the Elephantidae from the gomphothere assemblage reflects a major adaptive shift in the method of chewing and it was this new adaptation that seems to have accounted for the rapid expansion of the

---

[2] The semiaquatic moeritheres and the bizarre barytheres and deinotheres should probably not be included in the Proboscidea *sensu stricto,* although their exact relationships remain unknown.

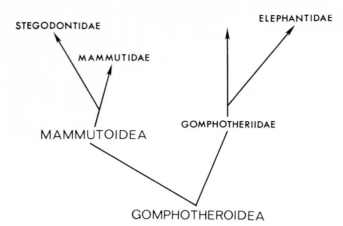

Fig. 1. Phyletic relations between the families of the Order Proboscidea.

family (sect. VII). Soon after their first appearance, the elephants replaced most of the gomphothere groups, with the exception of the Anancinae, and dominated the Pleistocene faunas of Africa, Eurasia, and North America. The new adaptive zone into which the elephants expanded provided the stimulus for a rate of evolution and success unknown in earlier Proboscidea. Within the family several phyletic groups evolved in parallel fashion, but their specific adaptations differed to the extent that their respective ecological requirements differed. These requirements in turn were subject to enormous variations as a result of wide geographic dispersal and temporal fluctuations in climatic conditions. At the present time little can be said concerning the ecological situations under which the various species lived, but this will be discussed further in a later section.

The potential value of the Elephantidae for stratigraphic correlation and for the study of evolutionary phenomena and zoogeography has long been appreciated. Yet, this potential failed to materialize in any meaningful way until recently because of the lack of knowledge relating to the earlier history of the family and because of insufficient stratigraphic and chronologic data. Recent discoveries of Pliocene deposits in Africa and a more firmly established geochronology based on better faunal resolution and absolute radiometric dating have greatly increased our understanding of this group (Cooke and Maglio, 1971). The record of elephantid history is now emerging as a valuable documentation of the evolutionary factors that led to their remarkable success. Although our knowledge is far from complete, and even far from adequate in many areas, there can be little doubt that today we are closer than ever before to realizing the potential value of the Elephantidae as a biological entity.

HISTORY

Because of the greater abundance of late Pleistocene deposits as compared with those of earlier ages, the discovery of fossil elephants began with the most recent and most specialized forms. Thus until 1823 the only fossil elephant known was *Mammuthus primigenius* from the late Pleistocene of Europe and North America. During the nineteenth century numerous other species from the middle and early Pleistocene were described and only during the last thirty years has the Pliocene history of the group become known. The latest Miocene-early Pliocene period in Africa, in which the earliest elephantines are represented, remained essentially unknown until 1965.

Partially because of this sequence of discovery, the relationships within the family have been confused. As earlier evolutionary stages were discovered or became better known, new phyletic interpretations were demanded. Thus, when only the woolly mammoth (*M. primigenius*) and the two living elephants were known, it was obvious that they merely represented variants on a common "elephant" plan. The significance of their differences was not apparent. Later, when the diversity of described species became almost overwhelming (e.g., see Osborn, 1942), a preoccupation with distinctness of forms seriously handicapped the study of evolution within the group.

Early attempts at reconstructing the phyletic history of the Elephantidae suffered from the nearly complete absence of paleontological data from sub-Saharan Africa. The earliest elephants in Eurasia were already well evolved and highly specialized forms. It was natural that their ancestry was sought among the only other proboscidean group with similar molar structure and geographic distribution, the stegodonts. Thus, among the more important phyletic reconstructions of their times were those of Leith Adams (1881), Weithofer (1888), Gaudry (1888), and Soergel (1912), all of whom considered the genus *Stegodon* as the ancestor of all later elephants (fig. 2). Both Weithofer and Soergel emphasized the distinctness of the African genus *Loxodonta* by suggesting an independent origin for this group. Such an interpretation was justifiable at that time on the evidence then available.

As earlier and more intermediate forms were discovered it became clear that *Stegodon* was too specialized to have given rise to the elephants and, as a result, the more primitive stegodont, *Stegolophodon,* was then considered the ancestral group (Osborn, 1936; Aguirre, 1969). Emphasis on dental criteria indicated a similarity between *Loxodonta* and several European species (e.g., *Elephas namadicus*) so that the African elephant was no longer considered unique. Despite these advances, however, direct evidence relating to the origin of the various lineages was still lacking. Reliance on only a few dental characteristics resulted in the grouping of unrelated species whose superficial resemblances were the result of retention of primitive characters or of parallel evolution.

It now seems almost certain that neither *Stegodon*

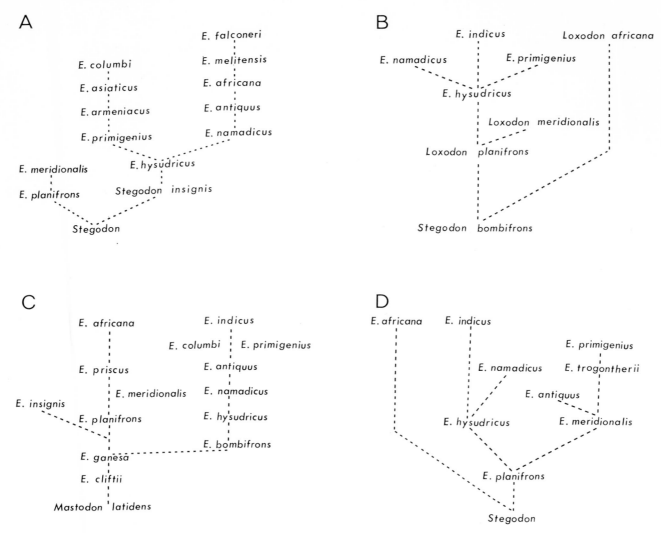

Fig. 2. Four early views on the phyletic relationships between species of elephants (*s.l.*); taken from, A: Leith Adams, 1881. B: Weithofer, 1888. C: Gaudry, 1888. D: Soergel, 1912.

nor *Stegolophodon* represents the group from which elephants were derived. Rather, the Gomphotheriidae can be traced through a transitional series of very primitive, hitherto unknown, elephants in Africa to the more progressive species of Europe and Asia (Maglio, 1970*a* and *b*; Maglio and Hendey, 1970).

SCOPE AND PURPOSE

The aim of the present study is to present a review of the available data on fossil elephants in order to clarify the evolutionary history of the group. In an attempt to establish a standard methodology for the analysis of elephant dental morphology, which has suffered in the past from inconsistency, a discussion of the meaning of taxonomic characters in this group is given. A standard method of measurement of these dental features is suggested in the hope that future workers will be able to make full use of the data presented here. It is also hoped that this (or another) standard approach to measurement will be adopted by all workers on fossil elephants so that published data can be more meaningfully compared.

In the systematic section an attempt is made to establish taxa of specific value that can be diagnosed and identified by other workers. This criterion has not always been applied in the past. Where nearly continuous lineages must be divided into manageable units for taxonomic convenience, small gaps in the record, or geographic criteria are utilized. In several cases where distinct segments of a lineage are recognizable but cannot be separated on any but arbitrary grounds, stages are designated, either with numbers or with locality names, in preference to formal subspecific terminology.

In order to evaluate the data in terms of phyletic relationships and evolutionary trends a provisional correlation of major Pliocene and Pleistocene deposits of the world is given. New potassium/argon age determinations and total faunal evidence (including elephants) were utilized in the construction of this chart. Based on this relative and absolute chronology, a discussion of the evolutionary events characteristic of the Elephantidae in general, and of each lineage in particular, is presented. A brief analysis of the rates of evolutionary change in dental and cranial morphology is also given as a comparison within and between the various phyletic lines.

The importance of the functional relationships between taxonomically significant characters is discussed in some detail along with a consideration of the function of the masticatory apparatus in elephants generally. The interplay of anatomical parts within this complex system must be understood if the evolution of the family is to make any sense in terms of specific adaptations.

Finally, the zoogeographic history of the family is discussed in an attempt to understand better the environmental and climatic gradients that may have influenced the evolution of the group, and to clarify the pattern of radiation and dispersal.

ABBREVIATIONS

The following abbreviations are used in the present work.

- AMNH The American Museum of Natural History, New York.
- BCUL Bedford College, University of London.
- BM British Museum (Natural History), London.
- CIT California Institute of Technology, Vertebrate Paleontology, Pasadena.
- CMD Colorado Museum of Natural History, Denver.
- DCR Dubois Collection, Rijksmuseum van Natuurlijke Historie, Leiden.
- FSA Faculté des Sciences, Algiers.
- GLM Geological Museum, Leiden.
- IGF Institute of Geology, University of Florence.
- IGR Geological Institute, Cittá Universitaria, Rome.
- IPUB Institute of Paleontology, Humboldt University, East Berlin.
- KNM National Museum of Kenya, Nairobi.
- MEO Museum of Education, Ochenanomizu, Tokyo.
- MMK McGregor Memorial Museum, Kimberley.
- MNHN Muséum national d'Histoire Naturelle, Paris.
- MNHT Museum of Natural History, Tripoli.
- MPP Museum of Paleontology, Pisa.
- NHB Naturhistorische Museum, Basel.
- NMB National Museum, Bloemfontein.
- PU Department of Geological and Geophysical Sciences, Princeton University.
- RMHL Rijksmuseum van Natuurlijke Historie, Leiden.
- SAM South African Museum, Cape Town.
- TMP Transvaal Museum, Pretoria.
- UCB Paleontological Museum, University of California, Berkeley.
- UNM University of Nebraska Museum, Lincoln.
- USNM United States National Museum, Smithsonian Institution, Washington.
- VJM Author's catalog reference for unnumbered specimens.

- ET Enamel thickness
- H Maximum crown height
- HI Hypsodonty index
- K/Ar Potassium/argon age determination
- L Overall length
- LF Lamellar frequency
- P Number of plates
- W Maximum width
- $\overline{M}$ Arithmetic mean
- M.Y. Million years
- N Number of specimens in a sample
- SD Standard deviation of the mean
- OR Observed range of the sample
- V Coefficient of variation

II. MORPHOLOGICAL CRITERIA

NATURE OF THE MORPHOLOGICAL CHARACTERS

The determination of meaningful and constant morphological criteria in dental and osteological features is fundamental to the recognition of discrete taxa and to the establishment of phyletic relationships. Besides strictly morphological analysis, stratigraphic, geographic, and, where available, paleoecological data often provide additional criteria upon which to base such relationships. Nevertheless, in many cases with vertebrate fossils, the paucity of adequate stratigraphic data increases our reliance on morphologic analysis which, therefore, must be cautiously carried out. In spite of this fact, minor characters often have been used to diagnose taxa or to establish relationships without adequate regard to their functional significance or to population variability.

From studies on the larger samples of fossil elephant specimens it becomes clear that certain characters such as enamel thickness and relative crown height provide valuable statistics upon which to base systematic studies. The degree of variability in these characters within known homogeneous assemblages, as in the living species, for example, shows them to be reliable in diagnosing taxa on the species level. Although they cannot *a priori* be taken as equally reliable for all fossil species, such an assumption is at least a good working hypothesis until additional evidence becomes available.

The vast majority of systematic studies on Proboscidea have been based on molar teeth, as these are the most common remains in the fossil record. Those measurements that seem most useful for identification are those that show progressive change along an evolving lineage. Although the gross evolutionary trends in the three phyletic lines here recognized are similar, they differ in detail and in rate of adaptive change so that equivalent segments of the different lineages (i.e., "species") can be diagnosed. The overall length, the number of plates, the height of the unworn crown and the thickness of enamel are the most important of these dental criteria. The last three characters show progressive change in every lineage, whereas the overall length varies irregularly, reflecting changes in absolute size of the animals.

Several ratios of these measurements contribute useful information. The most important of these is the relative crown height, or hypsodonty index. The length/width index, or relative crown width, is also often used, as is the lamellar frequency. The latter measures the number of plates in a standard distance of ten centimeters of crown length. Before using these measurements, however, it is necessary to know what they actually measure and what their limitations are. It is also necessary to establish the functional relationships between these measured parts and the overall molar structure.

Plate number: This is one of the most significant features of an elephant tooth in terms of functional requirements of mastication. As will be discussed below, the elephant molar is primarily a horizontal shearing device, not a grinding one. As such, it is the number of shearing surfaces crossing each other during the power or shearing stroke of the masticatory cycle that determines the volume of food broken down per stroke. Although it cannot be demonstrated that the throw of the mandible was similar in all elephants, from the skull architecture and presumed muscle configurations in fossil species this seems a reasonable assumption. In fact, it would have been mechanically impossible for the mandible of earlier species to have achieved a proportionately greater anteroposterior throw than either of the living species.

Thus, assuming a more or less equivalent jaw throw in most elephants, the number of plates per unit distance of molar length will be an important factor in the relative number of shearing edges available on the occlusal surface. The number of plates becomes an important functional feature of elephant molars, directly related to the shearing capacity of the tooth.

The relationship of plate number to functional shearing depends on concomitant changes in the overall size of the tooth. An increase in both length and plate number may result in an increase *or* decrease in the number of shearing edges per unit distance of jaw throw. It is the relationship between these two factors (the lamellar frequency) that determines the exact result.

Lamellar frequency: Because of the structural limitations of the molar and the overall size requirements determined by skull and mandibular architecture, changes in plate number require correlated changes in other structural features of the tooth. Thus, in order to maintain the molar at a reasonably constant length commensurate with limitations set by the maxilla and dentary, an increase in plate number requires an increase in packing, that is a decrease in absolute spacing of the plates. The lamellar frequency measures this absolute spacing in terms of the number of plates in a standard crown length of ten centimeters (fig. 3). The higher the lamellar frequency, the smaller is the absolute spacing between plates and the greater, therefore, the packing of the plates. Changes in the lamellar fre-

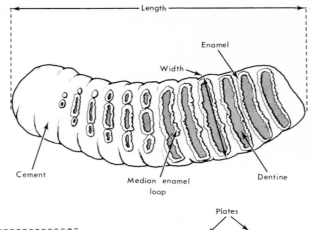

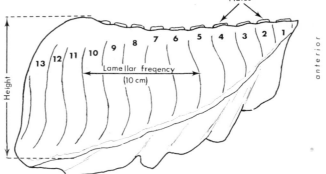

Fig. 3. Left M_2 of *Elephas namadicus* illustrating the measurable parameters and structural characteristics of elephant teeth as discussed in the text.

quency can reflect very different functional alterations of the molar structure and, therefore, must be considered in relation to other measurements. By itself, the value of the lamellar frequency often can be misleading.

For example, an increase in lamellar frequency from species A to species B indicates an absolute decrease in lamellar spacing along the tooth. If this was achieved through reduction in molar size without increase in plate number, the shearing ability of the tooth will have remained unchanged. However, if the increase in lamellar frequency reflects an increase in plate number, then it also reflects an increase in the functional status of the tooth.[3] A constant lamellar frequency can indicate no change in length and plate number and, therefore, no change in functional status of the tooth. It also may indicate a concomitant increase or decrease in both length and plate number so that absolute spacing remains constant. In the former case there is a functional change in which the number of shearing surfaces increases; in the the latter case the number of surfaces per unit stroke decreases.

---

[3] I avoid use of the term "efficiency" since this cannot be defined here in any meaningful biological sense. It is clear, however, that such changes in molar structure must reflect changes in functional ability—potential or actual.

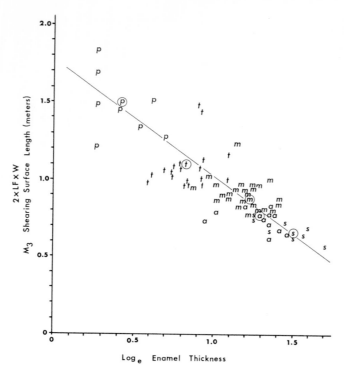

Fig. 4. Relationships between the $\log_e$ of the enamel thickness and the shearing index for M_3 of five species of *Mammuthus*. Total shearing index is defined as $2LF^3 \times 2LF_3 \times W_3/1000$ (see page 93), and measures the total length of shearing edge of enamel that functions in a standard jaw throw of 10 centimeters. The relationship shows a progressive increase in total shearing surface length with decrease in enamel thickness. Symbols indicate individual specimens as follows: s, *Mammuthus subplanifrons*. a, *M. africanavus*. m, *M. meridionalis*. t, *M. armeniacus*. p, *M. primigenius*. Circles indicate mean values for each species.

Thus, the significance of the lamellar frequency is in its relationship to changes in both plate number and molar length and, therefore, to the functional status of the structure. Alone, it may be useful as a comparative measurement between species, but in itself it has no functional significance.

Enamel thickness: This parameter is clearly related to the functional requirements of the elephant molar. As the number of plates increases and their spacing decreases, the enamel thickness is seen to decrease. This seems to be related to the structural limitations of the plate within the functional complex. As the plates become more closely spaced, they also become thinner so that cement intervals between plates do not become too narrow to be functionally useful. This thinning of the plate requires thinning of the enamel if the dentine interval within the plate is to be maintained. Without this enamel thinning the dentine (and cement) intervals between adjacent enamel surfaces would soon be reduced to insignificance. Thus, in order to retain a functionally viable structure of alternating soft surfaces that maintain shearing edges on the intervening hard enamel surfaces, the relative separation of enamel ridges must be maintained. The enamel must become progressively thinner.

The shearing function of worn elephant molars depends on the leading edges of enamel ridges as upper and lower molars pass over each other. It is the length of the enamel leading edges that is functionally significant in mastication, not its thickness. Thus, the thickness can be reduced, satisfying the functional needs of alternating cement-enamel-dentine structure, while at the same time the leading shearing edges of enamel can be increased (fig. 4).

A further modification of enamel structure is the degree of folding on its worn surface. Folding of the enamel increases the leading edge and, therefore, the total length of the shearing surface. The intensity of folding can be used as an indication of changes in functional status. Folding also increases the total thickness of the enamel ridge, providing additional strength to this surface.

Hypsodonty index: The relative crown height of an elephant molar is related to its ability to withstand wear. Thus, a molar with thick enamel subjected to soft vegetation can be lower crowned than one in which the enamel is thinner or which is subjected to more highly abrasive food. As the enamel thickness decreases in response to the increased packing of plates, the rate of occlusal wear increases greatly. Being softer than the enamel, the dentine and cement intervals wear down faster, leaving the resistant enamel ridges between them. It is, then, the rate of wear of these enamel ridges that determines the total rate of crown wear. As the enamel is reduced in thickness some compensatory change in other features of the tooth would be required to prevent rapid erosion. If diet were also changing to include more abrasive food, as was probably the case with the later mammoths, then these compensatory changes would have to be even more intense.

The major compensatory change related to reduction in enamel thickness is the increase in vertical height of the molar crown. This greater height allows a more rapid rate of wear without a reduction in molar life. There is, of course, no direct evidence that fossil elephants had a life span similar to that of the living species, but from their comparable sizes (except for the dwarfs) this would seem a reasonable assumption. The molar crown height differential between the most primitive and most progressive species may be as great as 400 per cent or more, yet both apparently wore down in approximately the same length of time. Thus, the rate of wear in the more progressive species would seem to be about four times as great as in the earlier form.

This relative crown height, or hypsodonty index, reflects the proportional molar shape by comparing height to width. This tends to eliminate size differences that would make it impossible to compare crown height in any functionally meaningful way. It is not the absolute height which is important here, but the relative height.

The potential height of the molar crown is limited by the structure of the skull and mandible, especially the latter. The limitations imposed by the mandible are greater than those of the maxilla primarily because of muscle attachment areas and the mobility of the jaw during mastication. As a result, the relative crown height of upper and lower molars is similar in early species, but in the later ones the upper teeth are considerably higher crowned. Further compensation is then needed to prevent the lowers from wearing out before the uppers. This is achieved by adding more plates to the lower molars and by altering the angle of shear of the enamel surfaces. These points will be discussed in more detail in Section VII.

Absolute size: The overall size of teeth is important for direct comparison between species and for use with the indices discussed above. The length of the crown is functionally related to the plate number and, therefore, to the shearing ability of the tooth. The crown width is an important factor in the length of shearing edges of enamel on the worn occlusal surface. The wider the tooth, the longer the shearing surfaces.

Length/width index: This is a measure of gross molar shape in terms of relative width. This proportion is often more important than either length or width alone and in some cases may be used as a specific distinction. In general, however, there is little difference in the value of this ratio from one species to another, and its usefulness is limited.

METRICAL PROCEDURE

The use of quantitative data for describing morphological characters is, in itself, of little biological interest except for comparative purposes and for the analysis of evolutionary trends. It is, therefore, essential that the same measurable features be utilized by all workers, or, less practically, that one worker examine all of the available material under study. Even when employing standard measurements, the ways in which each is taken can be as numerous as the investigators collecting the data.

In a study of molar variation in the living *Loxodonta africana,* Cooke (1947) lists seven standard measurements for elephant teeth and one ratio calculated from two of these. The convention of adding a plus sign (+) to the number of plates (P) was introduced to indicate an incomplete specimen. The method of taking these measurements was only briefly discussed so that the values given may be only approximately equivalent to those taken by other workers using different methods on the same material.

In an attempt to establish a standard set of measurements and a useful methodology for the collection of comparative data on elephant molar teeth, the methods employed in the present study will be discussed below. It is hoped that these will prove sufficiently meaningful and repeatable to be adopted by other workers, thereby providing a standard convention for future work.

P (plate number): A plate (often called lamella) consists of a single enamel fold with dentine filling its interior and with a "full sized" base at which it is joined to other such plates in front and behind. This last criterion is important since smaller anterior and posterior folds of enamel are often present. These appear as true plates on the occlusal surface when moderately worn, but they usually fuse into the base of adjacent plates in later stages of wear. It is conventional to place an X in front of or behind the plate number (P) to indicate the presence of significant folds that are not counted as full plates. As introduced by Cooke (1947), a plus sign should always be added to the plate number to indicate that the original number was greater. Individual plates should be referred to as P1, P2, etc. when counted from the front and PI, PII, etc. when counted from behind. If the root or root bases are present it is often possible to estimate accurately the number of missing plates on an incomplete specimen. When an estimated value is used in this or any other measurement, the superscript "e" is used after the given value.

L (length): Unlike teeth of other mammals those of elephants present special problems in the measurement of total length. A glance at the molars in figure 5 shows why this is so. If the measurement is taken parallel to the occlusal surface (A), which is often done, the result is a value considerably smaller than if it were taken parallel to the crown base (B). As a third alternative, the measurement could be taken perpendicular to the average lamellar plane (C). This latter method gives the most accurate measure of length since it is taken along the axis of growth and not along the variable planes of eruption or wear.

In a recent paper, Aguirre (1969) introduced a new measurement, the functional length, defined as the ratio between the number of plates in use and the length of the occlusal surface. The major problem with this calculation is that as the elephant molar erupts the angle at which the occlusal plane intersects the plates decreases, exposing a greater and greater number of plates per unit distance. In later stages of wear the relative spacing of plates will appear greater than in earlier stages. This phenomenon becomes more enhanced in more evolved species in which the molar crown height, and therefore the initial eruption angle, is greater. Although Aguirre's method does more accurately reflect the spacing of plates on the chewing surface at the time the animal lived, the dependence of this method on the degree of wear makes it less desirable as a comparative measurement. In the study of evolutionary phenomena within this group the genetic control of molar formation is more likely to be reflected in the basic

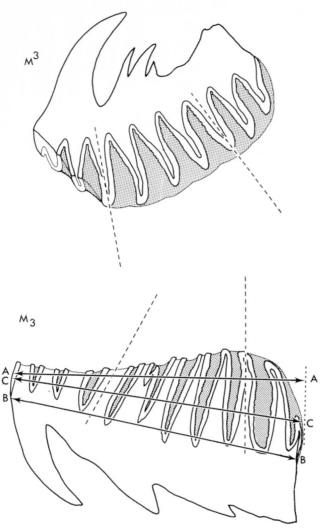

Fig. 5. Molars of *Elephas planifrons* in sagittal section showing the apically diverging plates on the upper, and apically converging plates on the lower molars of elephants. Planes of measurement for length are indicated as follows: A: parallel to occlusal surface. B: parallel to crown base. C: perpendicular to average plate direction. On higher crowned molars the discrepancies between these measurements would be greater. Shaded areas are cement.

preserved plate may not have been the widest on the original molar. The maximum preserved height may occur on a plate considerably posterior to the maximum width. By giving the plate number (as a superscript) of the given width, we can thereby place comparative measurements within lesser or greater limits of confidence.

H (height): The maximum crown height is measured vertically along a plate from the base of its enamel sheath to the highest point on its apex. This measurement should be taken parallel to the vertical axis of the plate which usually requires calipers with adjustable jaw lengths, since the highest point of the plate is near its mid-sagittal plane. When the plate is worn, a measurement of maximum preserved height can be taken along the side of the enamel and the value is then followed by a plus sign. As with the width, the maximum height is always given together with the plate number at which it was taken.

LF (lamellar frequency): This is a standard measurement of the average number of plates in a distance of ten centimeters along the anteroposterior axis of the tooth. The most accurate method of measurement is to calculate the LF from the number of true plates (excluding the terminal folds) and the length of this series of plates. Care must be taken to use the length only of the complete plates and equal number of cement intervals, since the LF should measure the number of plate-cement units within the ten centimeter interval.

More than any other measurement, the LF can show great variation depending on the worker or on the part of the tooth utilized. Generally, the plates of an upper molar diverge from the base of the crown so that the plates are more widely separated apically. The reverse situation prevails in lower molars. Thus, the LF will differ considerably from base to apex of the same molar. A second complicating factor is the lateral curvature of elephant molars. The uppers are concave inwards and the lowers are concave outwards. This results in a lower LF on the convex side (fig. 6).

Ideally, an accurate average measure of the LF can be calculated from the LF's taken at the base and apex on both the buccal and lingual sides of the molar. The mean of these four calculations should give the best value for the central portion of the molar. This is the method used here. When greatly worn, however, the value of the LF will obviously be distorted—too low for the upper, and too high for the lower molars.

Cooke (1947) introduced the value R (length-lamellae ratio = L/P) which measures the mean thickness of a complete plate-cement interval. If the LF is calculated as above, the value R becomes redundant, since it represents the reciprocal of the lamellar frequency and adds little additional information, although it does give a more direct measure of plate spacing. This calculation serves as a useful check on the lamellar frequency.

structure of the tooth. A standard method of plate spacing measurement is more reliable than one that is age dependent.

As in the case of plate number, a plus sign indicates an incomplete measurement and the superscript "e" an estimated value when associated with the length.

W (width): This simply measures the widest part of the molar (including cement). In order to make this measurement more useful when used in conjunction with other measurements (e.g., see hypsodonty index below), it is important to know on which plate the value was taken. In an incomplete specimen the widest

ET (enamel thickness): The enamel thickness is one of the most consistently reliable characters of the elephant molar. It is, therefore, important that standard methods of measurement be applied to this character. The most obvious source of error results from measuring the thickness parallel to the wear surface. Since the occlusal plane is inclined at an angle to the vertical axis of the plates, the worn enamel surface represents an oblique cut through the enamel sheath, making it appear thicker than it really is. Care must be taken to measure enamel thickness perpendicular to the outer face of one enamel surface, but not across a fold.

The actual thickness of enamel varies from one part of the plate to another, being generally thicker toward the apex and around the sides of the plate and on median loops. Aguirre (1969) suggested averaging a series of measurements along the crown. In addition, the maximum range of variation should be given for each molar.

HI (hypsodonty index): The relative measure of crown height was first used by Arambourg (1938) as an index for the hypsodonty of elephant molars. Later, Cooke (1947) modified this index by multiplying the height/width ratio by 100. This becomes a most valuable measure for comparing the relative crown height of different species, with a value of 100 indicating equivalence in height and width. The height and width are measured as above.

III. THE SPECIES AND GENUS CONCEPTS

Many pages have been printed concerning the problems of the species concept in paleobiology, but few conclusions have been reached that are applicable to all fossil material, or that are widely accepted. It is beyond the scope of the present discussion to review the numerous approaches applicable to the taxonomic units termed "species" by paleobiologists, or to attempt a comprehensive analysis of the problem. The purpose of this section is to specify the theoretical and practical approaches employed in the following pages so as to avoid confusion. At this point, it is more important to know the basis for taxonomic judgment than for all workers to agree on standard definitions. Although ideally desirable, the latter goal is still far from becoming a reality.

It has been argued that the "paleontological" concept of a species is based on arbitrary morphological criteria, whereas the neontological (or "biological") species concept is based on genetically assessed, discrete, non-arbitrary biological entities (e.g., Thomas, 1956). Yet, it is generally accepted that even the neobiologist relies, in most cases, on morphological criteria to infer genetic discontinuities (Mayr, 1963, 1970; Simpson, 1961). As emphasized by Simpson, once selected, morphological criteria do objectively define a taxon. It is the taxon, however, representing the taxonomist's concept of an assemblage, that is invariably subjective. Thus,

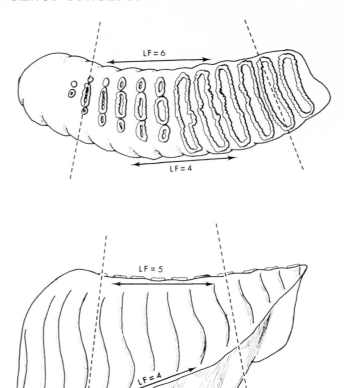

Fig. 6. Sources of error in measurements of lamellar frequency for elephant molars. These errors result from the non-parallel arrangement of the molar plates in both the vertical and horizontal planes. LF, lamellar frequency.

the use of morphological characters is, in itself, no reason to consider the species of the paleobiologist any different from that of the neobiologist.

The "biological" or "genetic" concept of species specifically requires genetic discontinuity, either actual or potential, and is tested by the criterion of interbreeding. Objection has been raised against this view because of the occurrence of interspecific hybrids between otherwise "good species." However, the concept of the transmutability of species through evolution demands a dynamic definition of species in which they pass through intermediate stages where genetic isolation is incomplete. As evolutionists we would worry if such intermediate stages did not exist. As Mayr (1963) and others have repeatedly emphasized, the existence of intermediate stages in no way destroys the concept of the species as a biologically definable, and therefore objective, entity.

For the paleobiologist, a problem arises in trying to apply this theoretical genetic concept of species to fossil groups. Simpson (*op. cit.*) attempted to broaden the genetic concept so as to apply to fossil assemblages by defining the species as a lineage evolving separately from others and with its own evolutionary role. The crucial point here is that the term interbreeding is not used and that the emphasis on distinction of evolutionary roles allows for incomplete genetic isolation. The definition does not object to the change of roles within species, but requires only that the role be distinct in some way throughout its duration. Simpson recognized the need for arbitrarily dividing continuous lineages which, under his definition, would otherwise have to be considered as single species. Thus, his definition does not help in defining segments of a single lineage, but merely gives a time dimension to the genetic concept which can be applied only to populations on different lineages.

As recently pointed out by Mayr (1970), the biological species concept has its primary significance with respect to sympatric and synchronous populations and is thus primarily non-dimensional. He goes on to say that the more distant two populations are in space and time, the more irrelevant biologically is their species status. With fossil populations on separate lineages, even when not exactly contemporaneous, the genetic implications of their distinct evolutionary histories are identical to those of living, contemporaneous species. However, we must agree with Mayr that as the temporal separation of these populations increases, the biological relevance of their genetic distinction decreases.

If we accept the genetic concept modified by Simpson's concept of distinct evolutionary roles, the definition of species can only be applied to one population in comparison to others which lie on different phyletic lines. Species so defined have objective, biological meaning. If two synchronous fossil populations are morphologically and ecologically distinct, we may infer that the genetic concept applies as it does for living animals, despite problems such as sibling species, which can never be tested for fossils anyway. Thus, when dealing with a small segment of two or more distinct lineages, some of the same criteria that apply to living populations can be utilized, for all living species are, by definition, on different phyletic lines.

In contrast to this, and as pointed out by other authors, successive stages in a phyletic continuum cannot be defined in any biologically objective way. Each segment represents a vague area without temporal, morphologic, or genetic boundaries (Jepsen, 1943). Although gaps in the fossil record resulting from non-preservation or non-discovery are certainly objective criteria in themselves, they have no biological meaning and cannot, therefore, be used to define adequately biological entities. "Species," in the sense of populations on genetically and evolutionarily distinct lineages, do not exist along a single evolving line. Nevertheless, when sufficient gaps are present which set off small segments of a lineage that can be defined on morphological, ecological and stratigraphic criteria, some recognition must be given to them. For convenience only, then, these segments can be treated as discontinuous, separate entities occurring on a common time plane and to which the genetic concept can then be applied as far as the evidence allows. But the biologically arbitrary nature of the resulting "species" must be kept in mind.

When new material on the same lineage becomes available from other localities, its distinctness from other known segments is assessed by the same criteria as used before. The important difference here is that these intralineage "successional species" are not equivalent to interlineage species. As intermediate populations are discovered, previously discontinuous segments of a lineage will become continuous and division on any but completely arbitrary criteria becomes impossible (Maglio, 1971). Although this has been pointed out by others, it has not always been applied by paleontologists (e.g., Crusafont-Pairo and Reguant, 1970).

The goal of the paleontologist is not the recognition of fossil species, but the establishment of evolutionary units. Therefore, the arbitrary division of more or less continuous lineages is counter to this goal. When gaps become too small, we should not hesitate to abandon previously defined segments (i.e., "species") in favor of a long sequence in which only subspecies or successive evolutionary stages can be recognized within a lineage The establishment of a continuous "species lineage," no matter how long, should be an acceptable exception to the practice of defining discrete units called "species." The major objective here is usefulness.

It is desirable that higher categories of classification, such as the genus, reflect levels of organization broader than that of the species, but which are narrow enough to indicate relationships. The species within a genus form a phylogenetic unit, descended from a common ancestor and functionally adapted to a similar mode of life. The genus as employed here is a group of one or more closely related phyletic lineages along which discontinuous segments or "species lineages" are recognized. All members of the genus share a fundamental adaptive zone different in some particular way from that of other such groups. This adaptive zone may or may not be precisely definable, but is inferred by the common possession of a particular suite of functionally integrated characters.

Because genera and other supraspecific taxa arise by the same micro-evolutionary processes responsible for the evolution of populations, and, therefore, first arise as new species, generic characters will be less distinct for earlier members of a genus. As still earlier and more transitional stages are found, generic distinctions will become more arbitrary, but this will not lessen the value of the genus defined on the entire assemblage.

IV. SYSTEMATIC REVISION

The term "systematics" in its broadest sense refers to the scientific study of the kinds and diversity of organisms and of the relationships among them (Simpson, 1961). In the present section it includes formal classification, nomenclature and morphological diagnoses with discussions of relationships restricted to forms of close affinity or similar structure. Broader patterns of phyletic relationship are discussed in a later section. Specific taxa are grouped within subfamilies and genera, but the groupings do not necessarily imply ancestral-descendant sequences. Rather, within each genus species are grouped generally by geographic occurrence; within these groupings lineages are arranged in ancestral-descendant sequences where possible.

An attempt has been made here to give a revised diagnosis for all taxa considered valid on current taxonomic usage. These diagnoses rely heavily on dental and cranial morphology where these are known. No attempt has been made to review post-cranial adaptations. Until a systematic examination of such elements is undertaken they are not likely to provide the type of data that are useful in a review of the kind presented here. This is not to say that skeletal elements lack taxonomic significance, for this is decidedly not the case. The problem lies in the fact that too little post-cranial material is presently available that can be definitely associated with dental remains. This is especially true for the earliest known species. Most of this material has yet to be described.

A classification of elephants was presented by Osborn (1936, 1942) in which numerous phyletic lineages extended backward in time without significant branching. This reflected Osborn's concept of non-reversability or non-alteration of established phyletic trends. In spite of this, however, Osborn's broad groupings were not greatly different from those proposed here. It is unfortunate that the criticism directed against that author's evolutionary ideas has prompted many workers to disregard many of his conclusions on proboscidean evolution. Osborn was able to deduce affinities between diverse groups that on present evidence seem more appropriate than some views proposed more recently.

A modified classification was presented by Simpson (1945) in which he reduced Osborn's five suborders to three. This scheme differed further in the grouping of both *Stegodon* and *Stegolophodon* with the true elephants, a notion originally held by Osborn but later abandoned when he transferred *Stegolophodon* to the Mastodontidae. The classifications of Simpson and Osborn are given below along with the scheme proposed in the present study. Taxa are listed to the family level except for the Elephantidae for which subfamilies and genera are also given. One obvious difference between the present scheme and all earlier ones is the suggested exclusion of the Barytherioidea, the Moeritherioidea, and the Deinotherioidea from the order. This proposal is not new here. Yet these groups have usually been included with the Proboscidea for convenience. In 1954 Deraniyagala (1954) first used the

CLASSIFICATIONS OF THE PROBOSCIDEA

Osborn, 1936	Simpson, 1945	Present scheme
Moeritherioidea	Moeritherioidea	
Moeritheriidae	Moeritheriidae	
	Barytherioidea	
	Barytheriidae	
Deinotherioidea	Deinotherioidea	
Deinotheriidae	Deinotheriidae	
Mastodontoidea	Elephantoidea	Mammutoidea
Mastodontidae	Mammutidae	Mammutidae
Bunomastodontidae	Gomphotheriidae	Stegodontidae
Humboldtidae	Elephantidae	Gomphotherioidea
Serridentidae	Stegodontinae	Gomphotheriidae
Stegodontoidea	*Stegolophodon*	Elephantidae
Stegodontidae	*Stegodon*	Stegotetrabelodontinae
Elephantoidea	Elephantinae	*Stegotetrabelodon*
Elephantidae	*Loxodonta*	Elephantinae
Loxodontinae	*Elephas*	*Primelephas*
Loxodonta	*Mammuthus*	*Loxodonta*
Palaeoloxodon		*Elephas*
Hesperoloxodon		*Mammuthus*
Elephantinae		
Elephas		
Hypselephas		
Platelephas		
Mammuthinae		
Archidiskodon		
Metarchidiskodon		
Parelephas		
Mammonteus		

new ordinal name Moeritheria for *Moeritherium* and *Barytherium*, but he did not offer adequate discussion. The distinctness of the latter genus had already been pointed out as early as 1906 when Andrews (1906) applied the name Barytheria[4] which Osborn (1936) raised to ordinal level. Recent study has suggested a closer relationship between *Deinotherium* and *Barytherium* than was previously believed (J. Harris, personal communication). The true affinities of the various groups will not be determined without new material and an extensive review of previous collections. Nevertheless, their inclusion in the Proboscidea is less than satisfactory even on present knowledge.

Another important difference between the present classification and those proposed by Osborn and Simpson is the implied relationship among the various proboscidean families. Osborn recognized three major groupings. The Mastodontoidea or "mastodonts" *sensu lato* included the mammutids, the gomphotheres, and *Stegolophodon*. The latter had been placed with the stegodonts and elephants in earlier works of Osborn, but he transferred it to the Mastodontoidea because of certain features of the dentition. The Elephantoidea included only the elephants while a new suborder, the Stegodontoidea, was erected for the stegodonts. Osborn envisaged the latter as an independent specialization paralleling the true elephants. As we shall see, this view was essentially correct. In Simpson's reclassification of 1945 the stegodonts and *Stegolophodon* were returned to the Elephantidae and were thought to have represented early transitional stages between mammutids and elephants. Thus Osborn's Elephantoidea was expanded to include the Mammutidae and Gomphotheriidae as well as the Elephantidae. This greatly altered his original (1921) concept of that taxon. Simp-

[4] This name is preoccupied by Barytheria Cope (1898) for the toxodonts.

son synonymized many of Osborn's generic names, retaining only *Loxodonta, Elephas,* and *Mammuthus* for the elephants.

In the classification proposed in the present work I have returned to some of Osborn's original views, although I have abandoned others. As was discussed elsewhere, *Stegolophodon* and *Stegodon* seem to represent a specialized line of mammutid descent, convergent on the elephant-type of masticatory pattern (Maglio and Hendey, 1970; Maglio, MS.). The Elephantidae were derived directly from the Gomphotheriidae. I have adopted a scheme in which the stegodonts and mammutids are grouped into a suborder Mammutoidea, and the gomphotheres and elephants into the Gomphotherioidea. The latter name replaces the name Elephantoidea because of the great disparity in usage between Osborn and Simpson and because of the predominance of the gomphotheres within this group. A new subfamily, Stegotetrabelodontinae, was previously established for the inclusion of *Stegotetrabelodon* (Aguirre, 1969). I have adopted Simpson's generic synonymy in most cases and have erected a new genus, *Primelephas,* for material unknown to him in 1945. The overall result is, I believe, a more simplified scheme in which evolutionary relationships are more accurately reflected.

Family ELEPHANTIDAE Gray, 1821

Elephantidae Gray, 1821: p. 305.
Elephasidae Lesson, 1842: p. 156.

Type genus: *Elephas* Linnaeus, 1758.
Included subfamilies:
  Stegotetrabelodontinae Aguirre, 1969.
  Elephantinae Gray, 1821.

Distribution: latest Miocene to Recent of Africa; late Pliocene to Recent of Eurasia; middle Pleistocene to sub-Recent of North America.

Revised diagnosis: Proboscidea with low- to high-crowned molar teeth bearing transverse ridges or plates, usually invested with cement; posterior columns in transverse valleys present in earlier species; valleys V- to U-shaped in cross section at the base, never Y-shaped as in stegodonts; premolars retained in earlier species. No more than one to one-and-a-half molars in use simultaneously in adults. Mandibular tusks present only in early forms, reduced and lost in more progressive species. Skull with ventrally depressed palate and vertically oriented alisphenoid wing; basicranium foreshortened; glenoid surface strongly rounded with a deep post-glenoid depression.

Subfamily STEGOTETRABELODONTINAE Aguirre, 1969

Stegotetrabelodontinae Aguirre, 1969: p. 1370.

Type genus: *Stegotetrabelodon* Petrocchi, 1941: p. 107.

Distribution: Latest Miocene to early Pliocene of Africa.

Revised diagnosis: Large elephants with low-crowned molars bearing few plates; M3 with six to seven transverse plates each superficially divided into columns by vertical grooves; strong median cleft not extending to the base of the crown; complete enamel loops forming with moderate wear; prominent isolated accessory columns behind the plates; crown height less than the width; enamel 4–7 mm. in thickness, not folded; transverse valleys between plates V-shaped in cross section but not compressed at the base; lamellar frequency 2.5 to 3.0. Mandible with long, protruding symphysis, usually with one pair of incisors present.

Stegotetrabelodon Petrocchi, 1941

Type-species: *Stegotetrabelodon syrticus* Petrocchi, 1941: p. 110.
Included species: The type-species; *Stegotetrabelodon orbus* Maglio, 1970b: p. 5.
Distribution: Latest Miocene to early Pliocene of North, Central, and East Africa. The major localities are Sahabi in southeastern Libya, Lake Chad, and Lothagam Hill (Lothagam 1), northwestern Kenya.
Diagnosis: As for the subfamily.
Discussion: The genus was first proposed by Petrocchi (1941) as a new proboscidean related to, but more progressive than, the most advanced known gomphotheres at that time. Its systematic position was left open and the genus was presumably to be included within the Gomphotheriidae. Little further attention was paid to it until Aguirre (1969) and Maglio (1970a) recognized the elephant affinities of the type species. The genus was subsequently transferred to the Elephantidae and made the nominate genus of its own subfamily.

The morphological distinctions of the stegotetrabelodonts, with their peculiar combination of relatively progressive elephantine characters and less progressive gomphotherelike features, clearly demonstrate their intermediate position in the transition from Gomphotheriidae to Elephantidae. In overall structure they are considerably more elephantlike and it is for this reason that the group is included within the latter family.

The low number of plates (6 to 7) in the third molars is just equivalent to the greatest number of cone-pairs (see page 87) attained in the most progressive gomphotheres of the genus *Anancus,* but structurally the dentitions of these two groups are quite divergent. The lingual and buccal cones in *Stegotetrabelodon* are equally subdivided into two parts each, and antero-posteriorly compressed to form a series of plates of the elephant type. Although a median cleft divides the lingual and buccal halves of the molar, as in gomphotheres, here it is present only on the apical half of the crown and is tightly compressed in the midline. Lingual and buccal portions of each plate are transversely in line, not offset as in *Anancus*.

The enamel columns and ridges which develop in profusion on molars of advanced gomphotheres are reduced in number and size in the stegotetrabelodonts, and only a single posterior column persists on the posterior face of the plates. In later species even these tend to become incorporated into the plates.

The crown height is not extensive (HI = 60 to 70), about equal to that in the latest gomphotheres and earliest elephantines. From this stage onward, however, the relative crown height becomes an important taxonomic character in the Elephantidae, becoming progressively greater in all lineages.

The structure of the transverse valleys remains primitive in being V-shaped at the base, but it is never compressed and Y-shaped as in stegodonts. When worn, the enamel surfaces of adjacent plates abut, thus reducing their effective shearing ability. The plates are massive in construction and widely separated.

A major difference between anancine gomphotheres and stegotetrabelodonts is the presence of an elongated mandibular symphysis in the latter, usually with long tusks.

Geographically, the genus extends from Libya to Kenya and westward into Uganda. Several poorly known localities in central Kenya demonstrate its presence there as early as 7.0 m.y. ago. The latest occurrence of the genus is questionably recorded from the Kanapoi beds at Ekora, dated at about 4.0 m.y. Although the evidence is far from adequate, it would appear that the stegotetrabelodonts were a widespread, and probably dominant, proboscidean group during much of the African Pliocene. There is no evidence to indicate that the genus ever extended its range outside that continent.

At present there are two known species, *S. syrticus* and *S. orbus*, which differ in a number of striking ways. A third, undescribed species from late Pliocene beds in the Lake Chad sequence (Coppens, pers. comm.) is apparently very distinct in a number of features.[5]

Stegotetrabelodon syrticus Petrocchi, 1941 [6]
(Pl. I, figs. 1–3)

Stegotetrabelodon syrticus Petrocchi, 1941: p. 110; 1953–1954: p. 11, pls. 1–4.[7]

Stegotetrabelodon lybicus Petrocchi, 1941: p. 107; 1953–1954: p. 41.
Stegolophodon sahabianus Petrocchi, 1943: p. 123, figs. 68, 69; 1953–1954: p. 45.

Holotype: MNHT (VJM SSLM 301), incomplete skull and jaw with worn left and right M3 in place; various skeletal elements.[8]

Hypodigm: The type and four additional third molars.

Types of synonyms: *S. lybicus*, IGR (VJM SSLM 302), right lower M3. *Stegolophodon sahabianus*, IGR (VJM SSUM 302), right upper M3.

Revised diagnosis: A large stegotetrabelodont with relatively slender molar plates and intermediate columns in every valley; relative crown height two thirds or less of the width; deep median cleft and widely spaced apical digitations on M3. Mandible with very long and slender incisors, closely spaced and forming about 57 per cent of the total jaw length; symphysis long, massive, and downturned at about 45 degrees to the corpus.

Discussion: The type molars are rather well worn so that comparison with unworn specimens is difficult without an adequate sample. The type of *Stegotetrabelodon lybicus* consists of a complete, unworn molar from the Sahabi beds. As discussed earlier (Maglio, 1970b), there is no reason to consider this as distinct from the type of *S. syrticus*. In all comparable morphological characters the two specimens are so similar that specific separation cannot be maintained. The likelihood of two very similar species from the same deposit, with no evidence for geographic or temporal isolation, is remote. Table 1 gives comparative measurements for the type specimens of these two species.

Similarly, *Stegolophodon sahabianus* Petrocchi (1943), also from the Sahabi beds, must be regarded as a synonym of *Stegotetrabelodon syrticus*. As with *S. lybicus*, the type of *sahabianus* consists of a single unworn molar, in this case an upper M3. It is distinctly progressive in appearance when compared with *S. syrticus*, but this is undoubtedly the result of its unworn condition. All the Sahabi molars as a group show no more variation between them than is seen in homogeneous collections of early elephants from other localities. Prominent intravalley columns and deep clefts are not present in the Asiatic genus *Stegolophodon*, but are present in *Stegotetrabelodon*. Although the type of *S. sahabianus* is larger than the few known specimens of *S. syrticus*, it is, nevertheless, within the expected range of variability as deduced from samples of *S. orbus* from Kenya. In table 1, the type of *Stegolophodon sahabianus* is compared with several specimens of *S. syrticus*.

With only three species described in the genus it is

---

[5] This material was recently described by Coppens, 1972, *C. R., Acad. Sci. Paris*, 274, p. 2964, as *Stegodibelodon schneideri*, but it does not appear to differ generically from *Stegotetrabelodon*.

[6] After this monograph was in press it was realized that the name *S. syrticus* was mistakenly chosen as the valid name for the present taxon since *S. lybicus* has page priority. Although synonymy cannot definitely be established, it seems likely that *lybicus* will prove to be the valid name for this species. This in no way alters the conclusions made here on the relationships of this taxon.

[7] In this and all similar lists only true synonyms not valid for other taxa are given for all taxonomic levels except the species, for which all other names applied to the various taxa are listed. To avoid lengthy lists of references, however, only the major work for each synonym is cited.

[8] Although skeletal remains are known for many fossil elephant species, no systematic study has ever been undertaken. They are, therefore, not discussed further in the present work.

TABLE 1

COMPARATIVE MEASUREMENTS FOR M3 OF *Stegotetrabelodon syrticus*, *S. lybicus* AND *S. sahabianus* FROM SAHABI, IN MILLIMETERS

Superscripts on measurements here and on all following tables denote the plate number on which the value was taken.

	P	L	W	H	LF	ET	HI
Upper M3							
S. syrticus MNHL, Type	6X	232.0	109.8³	73.0³	3.0	—	67.0
S. syrticus MNHL, ref.	—	—	119.0	80.0	2.8	7.1(6.9–7.2)	67.2
S. sahabianus IGR, Type	6X	242.0	126.1⁴	80.1⁴	3.2	6.0(5.3–6.3)	65.6
Lower M3							
S. syrticus IGR, Type	7X	280.0	115.0²	54.0⁶	2.9	5.8(5.5–6.0)	—
S. lybicus IGR, Type	7X	317.4	123.4³	74.1³	2.6	6.0	60.1

not possible to explore the evolutionary trends within the group. However, with respect to the kinds of changes known to have occurred during the transition from gomphotheres to elephants, *S. syrticus* must be regarded as the most "primitive" species, or the least elephantinelike of the three. The intravalley columns persist behind every plate and show no sign of the trend toward incorporation by fusion into the plate faces that occurs in later species. The crown is less consolidated with all clefts prominent and less tightly compressed than in *S. orbus* or the most primitive Elephantinae.

In the mandible of *S. syrticus*, the long symphysis bears a very long pair of parallel tusks that form 57 per cent of the total jaw length. The strong downward inclination of these incisors suggests that the head must have been carried high; this is also suggested by the relatively low position of the occipital condyles.

The only known skull of this species is crushed and lacks the portion above the level of the zygomatic arches. A reconstruction (fig. 7b), based on available data, suggests a more or less rounded skull with a nearly vertical occipital plane and downturned premaxillaries. The architectural modifications seen in later species of elephants, involving the forward displacement of the temporal region and consequent elevation of the occipital plane, plus the downward and backward deflection of the premaxillary and maxillary regions, are already well advanced in *S. syrticus* and unlike the condition in typical Miocene gomphotheres (fig. 7a). Possible functional adaptations for these changes are discussed in section VII.

Stegotetrabelodon orbus Maglio, 1970
(Pl. I, figs. 4, 5; pl. II)

Stegotetrabelodon orbus Maglio, 1970b: p. 5, pls. 1, 2.

Holotype: KNM LT 354,[9] complete left lower jaw with unworn left M_3, worn left M_2 and left and right incisors in place, incomplete left and right M^3, complete right M^2 and skeletal fragments.

Hypodigm: The type and twenty-seven additional specimens including various skeletal elements from the type locality. The best material is as follows: KNM LT 359, complete left and right M_3, left M^3 and right M^2 from a single individual. KNM LT 349, complete right M_3. KNM LT 352, complete left M_3. KNM LT 367, complete right M^3. KNM LT 342, complete left M_2 and partial right M_2. KNM LT 366, complete left M^2. KNM LT 365, complete right dm_3. Four additional specimens from the Kaperyon and Mpesida beds, Baringo, Kenya as follows: BCUL 2/31A, fragmentary left M_2. BCUL A2/36, ? left P^4. BCUL MP A/510, fragmentary left and right M_3. BCUL MP A/582, ? right P^4.

Horizon and locality: Late Miocene to early Pliocene, Lothagam 1, members b and c, Lothagam Hill, Kenya. Two specimens from the Mpesida beds and two from the Kaperyon beds, both in the Lake Baringo area of Kenya.

Diagnosis: About 12 per cent smaller than *S. syrticus*. Mandibular incisors short, forming about 38 per cent of the total jaw length. Free columns persisting only behind the first two plates on M3; relative crown height of M3 13 per cent greater than in *S. syricus*. Plate formula:

$$M3 \frac{6X}{7X}, \quad M2 \frac{5X}{5-6X}, \quad dm4 \frac{6}{?}, \quad dm3 \frac{?}{3}.$$

Discussion: With regard to most dental characters, *S. orbus* is not strikingly different from *S. syrticus* except in being smaller and, in general, more progressive

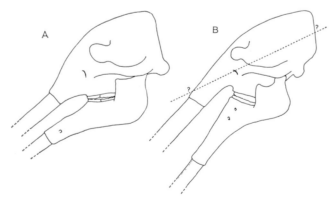

FIG. 7. Comparison of the skull of a Miocene gomphothere and that of the earliest elephant. Left-lateral view. A: *Gomphotherium angustidens*, after Gaudry, 1878; fig. 226. B: *Stegotetrabelodon syrticus*, new reconstruction, in part from Petrocchi, 1953–1954. Broken line indicates the level above which the cranium is reconstructed.

---

[9] Since the original description of the Lothagam and Kanapoi elephants (Maglio, 1970b), the locality code designations of catalog numbers have been changed from PAL LOTH, PAL KANAP and PAL EKA to KNM LT, KNM KP and KNM EK. All specimen numbers remain the same.

in the midline. The median loops formed by the vestigial columns remained, forming the typical lozenge-shaped pattern of this genus. Thus, far from being primitive in a structural sense, the Recent *Loxodonta* molar is a highly specialized structure, even though it has changed relatively less from the earlier pattern than have other genera.

The superficial resemblance to this dental pattern of molars in early species of *Elephas* (e.g., *Elephas namadicus* and *E. recki*) is not due to the median broadening of the plates as in *Loxodonta,* but primarily to the presence of a partially fused column which, when worn, gives a vaguely lozengelike pattern. When cranial evidence is used in conjunction with the dental morphology, it becomes clear that these two species of *Elephas* cannot be included in the genus *Loxodonta*.

In a recent description of a new proboscidean collection from near Lake Chad, Coppens (1965: p. 355) discussed several molar characters that he considered diagnostic for the genus *Loxodonta*. Among these were: (1) the tendency toward the development of a median loop by the persistence of the posterior column. But, as seen above, this is merely the retention of a primitive trait common to all early elephants and is not specific to this genus. (2) Smooth, unfolded enamel on worn molars. However, in molars of *L. atlantica,* coarsely folded enamel is a diagnostic feature. (3) The distinctive form of the molar cross section (not specified). The degree of variation he figures (1965; pl. 8, figs. 1–7), however, is so great as to render this almost meaningless as a diagnostic trait.

When considering molars of *L. atlantica* or *L. africana* there is little problem in recognizing the *Loxodonta* tooth form. But with early species, only cranial evidence provides certain identification. Nevertheless, molars of the most primitive stages of the Pliocene form *L. adaurora* with their very widely spaced plates, large posterior enamel loops, thick enamel, and low crowns are often separable from contemporaneous material which, on other characters, show clear *Elephas* affinities. Such identification may not be so easy for more evolved stages of *L. adaurora* and these may be difficult to separate from early stages of other groups.

Loxodonta adaurora Maglio, 1970
(Pls. V, VI)

Loxodonta adaurora Maglio, 1970b: p. 12, pl. 5, figs. 14–16, pl. 6, figs. 17, 18.
Elephas exoptatus Dietrich, 1942 (in part): p. 72, figs. 43, 54, 57, 59, 67, 68, and 73.
Archidiskodon cf. *meridionalis*, MacInnes, 1942: p. 92, pl. 8, fig. 3.
Archidiskodon planifrons, MacInnes, 1942: p. 86, pl. 7, fig. 10; pl. 8, figs. 1, 2.
Archidiskodon africanavus, Cooke and Coryndon, 1970: pl. 1, figs. D, E.

Holotype: KNM KP 385, a nearly complete skeleton, the skull and jaws bearing partially worn last molars in place.

Hypodium: The type and eighty-one other specimens of which the following are outstanding: KNM LT 353, skull with left M^2–M^3, right M^3 and left tusk, from Lothagam 3. KNM KP 383, left and right M^3, worn left M^3 and partial left and right M_3 from Kanapoi. KNM KP 390, complete right M^3. KNM KP 407, complete right M_3. KNM KP 406, mandibular ramus with M_1–M_2. KNM ER 346, skull fragment with left and right M^2–M^3, mandible with left and right M_3, from Kubi Algi. BM KE 10, left M_3 from Kanam East. BM M-15416, right M^1 from Vogel River. IPUB Vo. 330(7.78), left dm_4 from Vogel River. KNM YS (1967/21), right dm_3 from Yellow Sands (Mursi Formation). Numerous postcranial remains. The hypodigm includes the twenty-four syntypes of "*Archidiskodon exoptatus*" formerly referred to *Loxodonta* sp. (Maglio, 1969).

Horizon and localities: Early Pliocene to early Pleistocene of Africa. The type is from the Kanapoi Formation, southern Turkana, Kenya. Other specimens from the overlying Ekora Formation, Turkana, Kenya; unit 3, Lothagam Hill, Kenya; Bugoma, Kaiso Formation, Uganda; Kanam East and West, Kavirondo Gulf, Kenya; Yellow Sands (Mursi Formation), southern Ethiopia; the Vogel River Series, southern Serengeti, Tanzania; locality J.M. 514, Chemeron Formation, Baringo, Kenya; Kubi Algi and Koobi Fora, East Rudolf, Kenya; Chiwondo Beds, Malawi.

Diagnosis: Molars low-crowned, height equal to or less than the width; enamel not folded, 3 to 5 mm. in thickness; very large anterior and posterior columns, partially fused into the plates but free at their apices, forming prominent median loops with wear; plates thick and well separated, lamellar frequency from 2.6 to 4.4. Skull similar to that of *L. africana* but premaxillaries very large and somewhat flaring; frontal bones long with prominent fronto-parietal ridges lateral to external naris; occipital condyles high, projecting, but relatively small. Tusks long, massive and gently curved nearly in a single plane. Mandible with vestigial incisive germ cavities. Plate formula:

$$M3 \frac{8-10X}{10-11X}, \quad M2 \frac{7-8X}{6-8X}, \quad M1 \frac{7}{6-7},$$

$$dm4 \frac{5}{5-6}, \quad dm3 \frac{5}{5-6}, \quad dm2 \frac{3}{3}.$$

Discussion: Until recently, elephant molars here referred to *L. adaurora* have been described under a variety of different names, being grouped with species having broadly similar teeth. Thus, MacInnes (1942), in his discussion of elephant remains from Kanam and other East African localities, assigned material to two primarily non-African taxa. A partially worn M_3 (BM KE 10) was referred to *Archidiskodon* cf. *meridionalis*. From the degree of wear and the height of the last few plates, the hypsodonty index was prob-

TABLE 7
COMPARATIVE MEASUREMENTS FOR M3 OF *Loxodonta adaurora* MAGLIO 1970 FROM
KANAPOI AND OTHER AFRICAN LOCALITIES, IN MILLIMETERS

	P	L	W	H	LF	ET	HI
Upper M^3							
KANAPOI							
KNM KP 385 (type)	9X	245.1	108.0^3	91.8^8	3.9	4.1 (3.8–4.3)	85.0+
KNM KP 383	9X	291.5	120.5^6	109.2^4	3.0	3.6 (3.1–4.0)	99.5
LOTHAGAM							
KNM LT 383	10X	268.9	107.2^3	106.7^4	3.9	4.3 (3.7–4.8)	90.7
VOGEL RIVER							
IPUB KL 2/39	+5X	125.9+	82.7^{IV}	80.0^{III}	3.9	3.7 (3.3–4.1)	96.7
CHEMERON							
KNM 068/6409	4+	103.0+	93.0^4	66.0^4	3.7	3.2	71.0
KUBI ALGI							
KNM ER 346	9X	221.0	98.0^2	75.0^5	3.8	3.6 (3.2–4.0)	76.5
Lower M$_3$							
KANAPOI							
KNM KP 407	10X	303.1	105.3^5	95.0^5	3.5	4.2 (3.8–4.6)	90.2
KNM KP 385	11X	310.3	104.1^6	103.0^7	3.2	3.6 (3.0–4.3)	98.9
CHEMERON							
KNM 068/403	10^e	210.0+	90.0^6	92.0^6	3.8	3.0	102.1
KANAM EAST							
BM M15417	8+	193.4+	83.5^4	85.5^6	3.6	3.3 (3.0–3.6)	102.3
YELLOW SANDS							
Omo Coll. 67/23	+4X	126.0+	89.0	—	4.1	3.5 (3.0–4.0)	—
KUBI ALGI							
KNM ER 346	10X	249.0^e	89.0^4	82.2^5	3.8	3.9 (3.5–4.3)	92.2

ably not greater than 105 at the most. The lamellar frequency is 3.6 and the enamel thickness, 3.0 to 3.6 mm. The maximum number of plates was not over ten as judged from the position of the root bases. Although the enamel thickness is within the range of variation for *M. meridionalis,* both the LF and the number of plates are at the very lowest end of the observed range for that well-known species. The only occurrence of *M. meridionalis* in Africa is at Aïn Hanech, Algeria, provisionally correlated at less than 2 m.y. on faunal evidence. Both morphologically and chronologically the Kanam material seems clearly referable to *Loxodonta adaurora* (table 7).

The existence of *Elephas planifrons* in Africa has been argued by a number of authors (MacInnes, 1942; Arambourg, 1947; Hooijer, 1955; Coppens, 1965) on the basis of resemblances between molars from East Africa and from the Siwalik Hills, India. Hooijer (1955) convincingly demonstrated that molars of this type could be referred to the Asiatic species even though certain consistent differences distinguished the various geographic samples. As will be shown in a later section, skull criteria leave no doubt as to the distinctness of species formerly confused on dental evidence alone. The very typically *Loxodonta*-like skull structure of *L. adaurora* from Kanapoi differs markedly from that of *Elephas planifrons* from India.

Arambourg (1952: p. 407) described, as *Elephas africanavus,* a species from the late Pliocene of Lake Ichkeul, Tunisia. It was diagnosed on several features such as the strong lateral tapering of the molar plates, but these teeth do not differ fundamentally from East African ones. However, a poorly preserved skull from Garet et Tir, Algeria has recently been described by Arambourg (1969–1970) and this clearly shows strong affinities to species of the *Mammuthus* lineage. The molars on this specimen are somewhat less progressive than is typical for Lake Ichkeul material of *M. africanavus,* but it seems reasonably certain that the skull belongs to that species.

On molar evidence alone, then, it is difficult to separate adequately *L. adaurora* and *M. africanavus,* but again, in cranial osteology the two forms are clearly very different. The major distinctions are those used here as generic criteria and the two species are, therefore, placed in different major lineages.

A number of undescribed specimens from various East African localities should be referred to *L. adaurora* on present evidence (see table 7). A palate with left and right M^3 from the Ekora beds is close to the Kanapoi material and distinct from *Elephas ekorensis* which occurs in the same deposit. The number of plates is only nine and the enamel is smooth and thick (3.7 to 4.4 mm). The plates are rather widely separated with a LF of 2.6 to 2.7.

In a recent review of the "Laetolil" Proboscidea, it was shown that much of the syntype collection of *Archidiskodon exoptatus* Dietrich, 1942, was referable to *Elephas recki;* one of these specimens was chosen as the lectotype, thus reducing the name *"exoptatus"* to the synonymy of *E. recki* (Maglio, 1969). The remainder of the collection was referred to the Kanapoi species of *Loxodonta* (not named at that time). Although fragmentary, this Vogel River material is somewhat more progressive than the Kanapo sample in LF (3.4 to 4.3 vs. 2.5 to 3.9 in the Kanapoi collection), but is identical to it in all other respects. As far as molar teeth alone can be used within this group for identification, the Vogel River material can only be referred to *L. adaurora* at this time.

Several specimens from Bugoma in the Kaiso Formation were referred by Cooke and Coryndon (1971) to *Loxodonta africanava*. The Kaiso collection is incomplete, but on the available data it must be considered as conspecific with the Kanapoi species.

The "Yellow Sands" (Mursi Formation) elephant collection will soon be described by Y. Coppens. From material presently known, however, this form is identical to *L. adaurora*. At higher stratigraphic levels in the Omo sequence, in bed G (dated at 1.93 m.y.), additional material referable to this species has recently been collected (F. C. Howell, pers. comm.). This is clearly the same taxon described by Arambourg (1947: p. 270) as *Elephas planifrons* from the Shungura Formation.

Three specimens from Koobi Fora are assigned to the present species. A K/Ar date of 2.6 m.y. is available for a tuff in the lower member of Koobi Fora formation. It is possible that when better known this stage of *L. adaurora* may prove to be different on cranial evidence, but until that time it must be considered the same as the Kanapoi stage (Maglio, 1971).

The most primitive specimen currently included in this species is a fragmentary skull and associated jaw from Kubi Algi, East Rudolf, Kenya. The skull fragment has the deep maxillary fossa characteristic of this species and the backwardly inclined alisphenoid typical of the genus. The mandible is of the *Loxodonta* type, contrasting sharply with that of *Elephas recki* from the nearby Koobi Fora formation of later age. The upper and lower molar teeth have only nine and ten plates respectively, and these are widely separated. Strong enamel loops are free at their apices on the first several plates. The enamel is thick and not folded. The presence of this specimen at Kubi Algi is important in determining the geologic range for the species. Radiometric ages of 3.8 m.y. and 4.5 m.y. have been obtained for tuffs immediately above and below the fossil-bearing beds. Taken with the occurrences of this species in beds at the Omo and at Koobi Fora, the time range for this taxon must be taken as at least two million years, with very little change in molar structure throughout this period.

Loxodonta atlantica (Pomel), 1879
(Pl. VII)

Elephas atlanticus Pomel, 1879: p. 51; 1895: p. 42, pl. 8, figs. 1, 2.

Elephas (Loxodon) zulu Scott, 1907: p. 259, pl. 17, fig. 6, pl. 18, fig. 1.
Palaeoloxodon atlanticus, Osborn, 1942: p. 1274, figs. 1135, 1136.
Elephas pomeli, Arambourg, 1952 (in part): p. 413, figs. 7, 8; pl. 1, fig. 4.

Lectotype: Right M_2 in Musée d'Oran, Algeria (currently in the Muséum national d'Histoire Naturelle, Paris). This specimen was one of the two cotypes figured by Pomel (1879) and was refigured as such by Osborn (1942). It is taken here as the lectotype of the species name.

Hypodigm: The lectotype and about fifty additional specimens from North Africa, thirty-five from South Africa, and two from the Omo Valley.

Type of synonym: *Elephas (Loxodon) zulu*: PU 11548, left and right M_3.

Horizon and localities: The type is from the middle Pleistocene deposits at Ternifine, Mascara, Algeria. Other specimens from middle to late Pleistocene beds at Carrière Sidi Abder Rahmane, Oued Constantine and Sablière de Palekao, Algeria; middle to late Pleistocene beds at Elandsfontein, Cape, and from near Port Durfort, Natal, South Africa (type of *E. zulu*); and from below tuff E of the late Pliocene Shungura Formation, Omo Valley, Ethiopia.

Revised diagnosis: A large *Loxodonta* exceeding in size the living *L. africana*. Molars narrow with few plates, 12 to 15 on M3; crown height twice the width; plates relatively thick and widely spaced, LF 3.4 to 5.4 for M3; enamel figures lozenge-shaped with very strong median loops separated from the plate at their base by sharp constriction; median loops prominent, usually bifurcated and Y-shaped; irregularly placed open enamel folds along worn enamel figures; enamel relatively thick. Skull rounded, typically *Loxodonta*-like in structure; tusks and sheaths small, not as flaring as in *L. adaurora*; tusks gently curved nearly in a single plane, occipital condyles very large. Plate formula:

$$\text{M3}\frac{12-14}{10-15}, \quad \text{M2}\frac{9-12}{11-12}, \quad \text{M1}\frac{8-9}{7-10},$$

$$\text{dm4}\frac{8-10}{7-10}, \quad \text{dm3}\frac{6}{5-6}, \quad \text{dm2}\frac{?}{4}.$$

Discussion: Since its description by Pomel (1879) there has been no disagreement on the distinctness of this species, although numerous differing opinions on its relationships have been put forth. Scott's (1907) South African species, *Elephas zulu*, was for many years the only known representative of this species from south of the Sahara. While recognizing the great similarity between *E. zulu* and "*E.*" *atlanticus*, Cooke (1947), in an early revision of the Proboscidea of southern Africa, maintained *E. zulu* as a distinct species because of slight differences in size and crown height. In a later revision (1960) he followed Arambourg's

TABLE 8
Summary Measurements for Molars of *Loxodonta adaurora* Maglio 1970, Based on the Type and All Referred Specimens, in Millimeters

	P	L	W	H	LF	ET	HI
M³ OR	8–10	228.1–302.2	76.0–124.9	66.0–114.0	2.6–4.0	3.2–4.4	71.0–107.2
$\bar{M}$	9.1	271.6	104.5	92.6	3.4	3.9	92.4
SD	0.6	25.7	13.6	15.6	0.4	0.3	11.0
N	8	8	16	12	15	16	11
V	7.0	9.5	13.0	16.8	13.3	8.1	11.9
M₃ OR	10–11	231.0–310.3	69.2–111.2	71.5–103.0	2.9–4.4	2.5–4.8	84.0–103.3
$\bar{M}$	10.1	272.6	90.2	86.2	3.7	3.8	95.6
SD	0.4	30.7	11.6	9.0	0.4	0.6	5.8
N	6	6	22	16	21	20	16
V	4.0	11.3	12.8	10.4	12.2	15.4	6.1
M² OR	7–8	—	75.9–100.0	70.0–88.0	3.3–4.9	3.0–4.4	90.7–98.3
$\bar{M}$	7.5	—	87.5	76.7	4.0	3.6	93.7
SD	—	—	9.8	9.8	0.6	0.5	4.0
N	2	—	6	3	6	6	3
V	—	—	11.2	12.8	15.5	13.8	4.3
M₂ OR	6–8	138.0–191.1	69.0–91.1	70.0–90.0	3.6–5.0	3.0–4.5	90.7–111.6
$\bar{M}$	7.0	165.5	82.6	81.0	4.3	3.8	101.8
SD	—	—	6.2	6.1	0.5	0.4	7.1
N	4	4	11	8	11	11	8
V	—	—	7.5	7.5	10.7	11.9	7.0
M¹ OR	7	194.1	85.0	68.0	3.5–4.3	3.4–3.9	80.0
$\bar{M}$	—	—	—	—	3.9	3.6	—
N	1	1	1	1	2	2	1
M₁ OR	6–7	152.0–165.0	69.3–85.8	74.7–75.0	4.6–5.1	2.8–3.8	87.7–108.2
M	6.3	156.5	75.7	74.9	4.8	3.5	98.1
SD	—	—	5.5	—	0.2	0.4	—
N	3	3	6	3	7	7	3
V	—	—	7.3	—	4.4	10.8	—
dm⁴ OR	5	—	47.2–66.8	52.3–62.0	4.0–7.3	2.2–3.0	92.8–110.8
$\bar{M}$	—	—	56.7	62.0	5.9	2.5	101.8
N	1	—	3	2	3	3	2
dm₄ OR	5–6	126.4	53.6–58.0	40.2	5.0–6.4	2.1–2.8	69.4
$\bar{M}$	5.5	—	56.8	—	5.8	2.4	—
N	2	1	4	1	4	4	1
dm³ OR	5	47.1–72.0	33.0–41.0	31.6–42.5	8.0–10.0	1.2–2.0	77.1–108.9
$\bar{M}$	5.0	63.4	38.5	38.8	8.6	1.7	96.3
N	3	3	4	3	4	4	3
dm₃ OR	5–6	54.0–73.2	34.2–39.8	28.6–32.0	7.1–10.0	1.5–2.0	79.2–84.6
$\bar{M}$	5.4	64.9	37.1	30.2	9.1	1.7	78.2
SD	0.5	8.0	2.2	1.7	0.9	0.2	5.3
N	7	7	7	4	7	6	4
V	9.8	12.4	5.9	5.6	10l3	11.5	6.7
dm² OR	3	20.3	15.5	15.3	—	—	98.8
$\bar{M}$	—	—	—	—	—	—	—
N	1	1	1	1	—	—	1
dm₂ OR	—	19.2–26.3	15.3–19.7	13.7–18.6	—	—	85.6–101.9
$\bar{M}$	3	22.3	17.0	16.0	—	—	94.0
N	3	3	3	3	—	—	3

TABLE 9
Comparative Measurements for Molars of *Loxodonta atlantica* (Pomel) 1879 from Various African Localities, in Millimeters

	P	L	W	H	LF	ET	HI
Upper M³							
TERNIFINE							
MNHN [no no.]	14X	327.8	89.3[5]	154.1[10]	4.4	3.4(3.0–3.7)	171.5
MNHN [no no.]	8+	246.0+	89.4[7]	188.0[5]	3.6	3.5(3.1–3.8)	211.5
ZULULAND							
PU 11548	12e	252.1	84.7[5]	84.9[10]	4.3	2.7(2.3–3.1)	117.9+
ELANDSFONTEIN							
SAM 2546	10+	184.0+	82.0[5]	149.0[7]	5.2	2.3(2.0–2.5)	173.9
SAM 2552	+8X	217.0+	102.0[2]	135.0[4]	4.6	2.5	137.8
Lower M₃							
TERNIFINE							
MNHN 55-15-594	14X	403.1	97.0[7]	146.5[11]	3.9	3.5(3.0–4.0)	150.5
MNHN, Type	15e	340.0e	79.2[6]	140.1[9]	3.9	3.6(3.2–3.9)	175.0
ELANDSFONTEIN							
SAM 2534	13X	327.0e	78.0[6]	145.0[9]	4.4	2.3(2.0–2.5)	202.8
SAM 9471	9+	203.0+	110.0[5]	162.0[9]	4.0	2.5	165.3
SAM 5283B-D	9+	207.0+	81.0[4]	140.0[8]	4.4	2.3(2.0–2.5)	173.0
Upper M²							
TERNIFINE							
MNHN 25-2	10	220.0	82.8[8]	125.1[8]	4.5	2.7(2.3–3.1)	154.2+
MNHN 55-13-339	10e	169.1+	75.2[4]	132.4[8]	4.4	3.1(2.9–3.2)	176.0
MNHN [no no.]	9	229.2	84.6[6]	134.0[8]	4.3	2.8(2.4–3.1)	169.8
ELANDSFONTEIN							
SAM 14220	12X	216.0	80.0[5]	106.0[8]	6.5	2.3(2.0–2.5)	139.5
SAM 16782	+9+	—	79.0[3]	111.0[5]	5.8	2.0	136.3
Lower M₂							
CONSTANTINE							
MNHN [no no.]	12e	258.6+	70.3[8]	122.6[5]	5.1	3.3(3.0–3.6)	183.5
MNHN 54-7-671	11X	227.3	68.9[5]	108.0[9]	5.1	2.7(2.3–3.0)	162.2
ELANDSFONTEIN							
SAM 11161	+6	142.0+	72.0[3]	131.0[6]	5.5	1.3(1.0–1.5)	182.2

(1938) earlier suggestion that Scott's species fell completely within the range of variability of *"E." atlanticus* from North Africa. As long as Scott's type molars were the only known specimens from southern Africa, little progress was possible on this issue, and not until the recently collected fauna from Elandsfontein became available could the northern and southern samples adequately be compared.

A number of consistent differences between these two samples are present which, though not great, may be used to define two discrete morphological units. In the Ternifine collection the median loop tends to be bifurcated when worn, forming a Y-shaped fold. These loops are single folds in the southern group. The North African collection tends to be smaller in size than the southern and the lamellar frequency tends to be slightly lower for upper molars. But the greatest difference is in the enamel thickness, which, in the South African material, is consistently and significantly thinner with little overlap in observed range with the northern populations (table 9). These distinctions are not sufficient to warrant specific separation, however, and I follow Arambourg and Cooke in considering the two groups as conspecific. Whether or not they should be recognized as subspecific taxa or stages is a matter of taxonomic preference. Although I do not feel that formal nomenclatural distinction would add appreciably to our understanding of this species or make it any easier to deal with it in taxonomic usage, it is probably best to maintain the two geographic assemblages for convenience in discussion. Accordingly, the following two subspecific designations may be applied:

Loxodonta atlantica atlantica (Pomel), 1879

Lectotype: As for the species.
Distribution: Middle Pleistocene of North Africa.
Diagnosis: A subspecies of *L. atlantica*, smaller than the southern form; plates on upper molars less closely spaced, LF 3.6 to 4.4; median loops on worn occlusal surfaces bifurcated and Y-shaped; enamel relatively thick, 3–4 mm.

Loxodonta atlantica zulu (Scott), 1907

Holotype: PU 11548, left and right M₃.
Distribution: Middle Pleistocene of South Africa.

TABLE 10

SUMMARY MEASUREMENTS FOR THE PERMANENT MOLARS (M3–M1) OF THE NORTH AFRICAN AND SOUTH AFRICAN POPULATIONS OF *Loxodonta atlantica* (POMEL) 1879, IN MILLIMETERS

	P	L	W	H	LF	ET	HI
North Africa							
M³ OR	14	327.8	84.0–98.0	154.1–188.0	3.6–5.2	3.4–3.5	171.5–211.5
M̄	—	—	90.6	171.0	4.4	3.4	191.5
N	1	1	3	3	3	3	2
M₃ OR	10–15	326.8–403.1	77.8–108.3	140.1–146.5	3.9–4.9	2.2–3.6	150.5–175.0
M̄	13.4 –	356.6	88.3	143.3	4.4	3.3	162.7
SD	1.9	—	13.7	—	0.5	0.3	—
N	5	3	5	2	5	5	2
V	14.5	—	15.5	—	11.5	8.8	—
M² OR	9–11	211.0–229.7	72.0–84.6	120.0–174.1	3.6–4.8	2.5–3.2	152.2–212.2
M̄	9.8	222.5	77.8	138.7	4.4	2.8	178.5
SD	0.8	—	4.4	18.4	0.4	0.2	24.2
N	6	4	9	7	9	9	7
V	7.7	—	5.7	13.3	8.3	7.5	13.6
M₂ OR	11–12	227.3–255.6	68.9–81.1	108.0–122.6	3.9–5.3	2.7–4.0	162.2–183.5
M̄	11.3	242.0	73.4	115.3	4.7	3.4	172.8
SD	—	—	5.7	—	0.6	0.5	—
N	3	3	5	2	5	5	2
V	—	—	7.8	—	13.3	14.6	—
M¹ OR	8–9	184.0–189.5	72.9–81.2	115.9–135.0	4.7–4.9	2.3–2.9	156.1–180.0
M̄	8.5	186.7	76.2	125.4	4.8	2.7	168.0
N	2	2	3	2	3	3	2
M₁ OR	7–8	180.0–188.5	65.0–75.1	—	3.7–4.8	2.2–3.4	—
M̄	7.5	184.2	69.6	—	4.1	2.7	—
N	2	4	4	—	4	4	—
South Africa							
M³ OR	—	252.1–263.3	82.0–102.0	135.0–189.0	4.0–5.4	2.0–2.8	173.9–206.0
M̄	12	257.7	88.3	157.7	4.7	2.5	189.9
SD	—	—	7.9	—	0.6	0.3	22.7
N	2	2	5	3	5	5	2
V	—	—	8.9	—	12.6	13.0	11.9
M₃ OR	13	227.0	78.0–110.0	131.0–172.0	3.4–4.5	2.0–2.8	135.7–202.8
M̄	—	—	89.5	152.2	4.1	2.4	172.9
SD	—	—	11.3	14.2	0.4	0.2	24.7
N	1	1	9	8	9	9	8
V	—	—	12.6	9.3	10.0	9.2	14.3
M² OR	12	216.0	79.0–80.0	106.0–111.0	5.8–6.5	2.0–2.3	136.3–139.5
M̄	—	—	79.5	108.5	6.1	2.1	137.9
N	1	1	2	2	2	2	2
M₂ OR	—	—	69.0–81.0	131.0	4.0–5.6	2.0–3.0	182.2
M̄	—	—	75.0	—	4.9	2.2	—
SD	—	—	4.6	—	0.7	0.4	—
N	—	—	5	1	5	5	1
V	—	—	6.2	—	13.3	19.4	—
M¹ OR	—	—	57.0–70.0	109.0	5.3–6.0	—	155.7
M̄	—	—	63.5	—	5.6	2.0	—
N	—	—	2	1	2	2	1
M₁ OR	10	—	66.0–67.0	—	4.9	2.0–2.8	—
M̄	—	—	66.5	—	—	2.4	—
N	1	—	2	—	1	1	—

Diagnosis: Larger than the northern forms of the species; plates on upper molars more closely spaced, LF 4.3 to 5.2; median enamel loops simple, not bifurcated; enamel relatively thin, 2–3 mm.

Arambourg (1952: p. 413) described a new species of elephant from North Africa as *Elephas pomeli* on the basis of three specimens. The type, from Carrière Sidi Abder Rahmane, is not distinct from *Elephas iolensis* Pomel and is included below in the synonymy of that species. The other specimens, from the Maison Carrée Formation, Oued Constantine, Algeria, plus more recently collected ones from the same locality, are referable to *Loxodonta atlantica atlantica*. On a sectioned specimen figured by Arambourg (1952: fig. 8) the wear figures clearly show the median expansion and large, Y-shape loop so typical of this taxon. The molars are narrower and the height lower than in *E. iolensis*.

During the 1969 field season of the American Expedition to the Omo Valley, a pair of upper third molars of an early stage of *L. atlantica* was recovered from below tuff E of the Shungura Formation. Chronologically, this level is considerably older than Ternifine and falls within the latest part of the Pliocene as defined here. More significantly, this is the only specimen of *L. atlantica* known from East Africa. These molars are identified as third or last molars on the basis of the narrowing of the last plate and the lack of a flat or concave posterior surface to the root base. If this is correct, then they are considerably more primitive than typical *L. atlantica*, as would be expected from their early age. The Omo specimens differ from those of the Ternifine and Elandsfontein collections in being smaller, in having fewer plates and in being considerably lower crowned. The last plate measures 71.9 mm. in height and must have been about 8 to 10 mm. greater when unworn. Allowing for a maximum height of about 110 mm. for the complete molar, the maximum estimated HI could not have been more than about 150 to 160. This is still below the observed range for all other M3's of this species. The enamel thickness is intermediate between that of the northern and southern samples. The subspecific status of this specimen is uncertain without a larger sample.

The occurrence in East Africa of this earliest and most primitive known stage of *Loxodonta atlantica* suggests a late Pliocene origin for the species, perhaps from *L. adaurora*. There are no morphological features in the latter that would preclude such an ancestry and the temporal relationship between the two forms lends support to the hypothesis. The very rare occurrence of this species at Omo, and its absence from other East African localities in which *Elephas recki* is abundant, and the absence of the latter from Ternifine and Elandsfontein, suggests that these two species were adapted to different ecological situations. This may account for the disappearance of *L. atlantica* in the lower Pleistocene of East Africa and its apparent replacement by *E. recki*.

Loxodonta africana (Blumenbach), 1797

Elephas africanus Blumenbach, 1797: p. 125, fig. C.
Elephas capensis G. Cuvier, 1798: p. 149; 1799, pl. 3, fig. 2.
Elephas priscus Goldfuss, 1823: p. 489, pl. 44.
Elephas cyclotis Matschie, 1900: p. 194.
Elephas oxyotis Matschie, 1900: p. 196.
Loxodonta africana Roosevelt & Heller, 1914: p. 739.
Loxodonta prima Dart, 1929: p. 725, figs. 25, 26.
Loxodonta atlantica angammensis Coppens, 1965: p. 360, pls. 11–13.

Holotype: Right M_2, whereabouts unknown.

Distribution: Type locality (Osborn, 1942: p. 1197) "probably Cape Colony, South Africa." Currently widespread in Africa south of the Sahara. Known as fossil in middle and late Pleistocene deposits of South Africa and Chad.

Discussion: Since this work is primarily concerned with fossil forms, no attempt has been made to examine each of the subspecies or types of synonyms listed above. For the most part I follow recent authorities. In 1907, Lydekker (1907: p. 384) suggested synonymy of *Elephas capensis* with *Loxodonta africana*. Later, Osborn (1924) formally reduced *E. capensis* to a subspecies of *L. africana* but did not diagnose it. Cuvier (1798) did not compare his material with Blumenbach's earlier type of *L. africana*, but it seems probable that Osborn was correct in regarding the two as conspecific. The type of Matschie's *Elephas cyclotis* is a skull from Mwelle district in the southern Cameroon. Although Osborn apparently regarded this form as no more than a subspecies of *L. africana*, when his monograph was published posthumously in 1942, an editorial note (p. 1196) directed special attention to a work by G. M. Allen (1936), not seen by Osborn, in which the Cameroon forest elephant is maintained as a valid species. Allen's reasons for specific recognition of this taxon were based on supposed ancestral-descendant relationships and on morphological differences. However, in the author's opinion the distinctions in ear form and size do not justify more than subspecific rank, if that.

The important question of whether *Loxodonta* was present in Pleistocene deposits of Europe rests on the determination of the type and referred specimens of *Elephas priscus* Goldfuss, 1823.[10] The type specimen was originally in the collection of Canon Mehring of Cologne. The locality from which it derived was unknown, but was assumed to be the neighborhood of Cologne. Cuvier (*in* Falconer, 1868: p. 95) doubted the authenticity of the type specimen as a fossil and considered it to belong to the living African elephant. Earlier, Falconer (1857, pl. 14, figs, 6, 7) had figured two referred specimens from Gray's Thurrock in the

---

[10] Although he figured these specimens in 1821, Goldfuss did not apply the name *E. priscus* to them until 1823.

TABLE 11
Summary Measurements for all Molars of *Loxodonta atlantica* (Pomel) 1879, North and South African Localities Combined, in Millimeters

	P	L	W	H	LF	ET	HI
M^3 OR	12–14	252.1–327.8	82.0–102.0	135.0–189.0	3.6–5.4	2.0–3.5	171.5–211.5
$\overline{M}$	12.7	281.1	89.2	164.3	4.6	2.8	190.7
SD	—	—	7.2	22.0	0.6	0.6	—
N	3	3	8	6	8	8	4
V	—	—	8.1	13.4	13.9	19.8	—
M_3 OR	10–15	326.8–403.1	77.8–110.0	131.0–172.0	3.4–4.9	2.0–3.6	135.7–202.8
$\overline{M}$	13.3	349.2	89.1	150.3	4.2	2.7	170.9
SD	1.8	—	11.7	13.1	0.4	0.5	23.0
N	6	4	14	10	14	14	10
V	13.1	—	13.1	8.7	10.7	18.8	13.4
M^2 OR	9–12	211.0–229.7	72.0–84.6	106.0–174.1	3.6–6.5	2.0–3.2	136.3–212.2
$\overline{M}$	10.1	221.2	78.1	132.0	4.7	2.7	169.5
SD	1.1	8.2	4.0	20.8	0.8	0.3	27.6
N	7	5	11	9	11	11	9
V	10.5	3.7	5.4	15.8	16.6	12.6	16.3
M_2 OR	11–12	227.3–258.6	68.9–81.8	78.0–131.0	3.9–5.6	2.0–4.0	162.2–183.5
$\overline{M}$	11.3	242.0	74.2	109.9	4.8	2.8	176.0
SD	—	—	5.0	—	0.6	0.8	—
N	3	3	10	4	10	10	3
V	—	—	6.7	—	12.7	27.2	—
M^1 OR	8–9	184.0–189.5	57.0–81.2	109.0–135.0	4.7–6.0	2.0–2.9	155.7–180.0
$\overline{M}$	8.5	186.7	71.1	120.0	5.1	2.4	163.9
SD	—	—	8.9	—	0.5	0.5	—
N	2	2	5	3	5	5	3
V	—	—	12.5	—	10.3	18.8	—
M_1 OR	7–10	180.5–188.5	65.0–75.1	—	3.7–4.9	2.0–3.4	—
$\overline{M}$	8.3	184.2	68.6	—	4.3	2.6	—
SD	—	—	3.8	—	0.6	0.5	—
N	3	2	6	—	5	6	—
V	—	—	5.6	—	13.1	19.3	—
dm^4 OR	8–10	137.0–185.0	48.0–66.2	109.5	4.9–8.6	1.0–2.7	140.0–181.8
$\overline{M}$	8.7	153.7	59.1	—	6.4	1.9	181.8
SD	—	—	7.0	—	1.4	0.7	29.6
N	4	3	5	1	5	5	2
V	—	—	11.8	—	22.5	35.3	18.4
dm_4 OR	7–10	146.0–194.9	42.0–60.5	88.0–111.0	4.9–9.5	1.0–2.8	161.8–191.0
$\overline{M}$	9.0	170.0	52.4	98.0	6.1	2.0	177.9
SD	1.2	17.4	5.7	—	1.5	0.5	—
N	5	5	9	3	9	9	3
V	13.6	10.3	11.0	—	23.9	23.5	—
dm^3 OR	6	60.0	29.5–38.0	35.0–36.4	9.7–10.1	1.0–1.2	110.3–118.8
$\overline{M}$	—	—	33.5	35.7	9.9	1.1	114.5
N	1	1	3	2	2	3	2
dm_3 OR	5–7	57.3–69.0	29.0–37.0	33.0–38.4	9.2–11.2	0.5–1.2	117.1–131.1
$\overline{M}$	6.1	62.5	33.1	36.1	11.6	1.0	122.8
SD	0.7	3.9	3.0	—	0.9	0.2	—
N	7	7	7	4	7	7	3
V	11.2	6.2	9.0	—	8.3	23.0	—
dm_2 OR	—	22.0–23.0	—	—	17.4–18.2	0.8–1.0	—
$\overline{M}$	4.0	22.5	14.0	—	17.8	0.9	—
N	2	2	2	—	2	2	—

TABLE 12

Summary Measurements for Molars of *Elephas ekorensis* Maglio 1970, in Millimeters

	P	L	W	H	LF	ET	HI
M³ OR	—	276.9–304.1	72.7–96.6	87.5–113.2	3.8–4.5	3.3–4.0	108.0–127.7
$\bar{M}$	11.0	290.5	87.1	100.6	4.0	3.7	115.2
SD	—	—	9.1	8.8	0.3	0.3	7.0
N	2	2	6	6	6	6	6
V	—	—	10.5	8.8	6.9	7.0	6.1
M₃ OR	—	280.0–285.0	87.6–96.5	95.4–111.9	4.0–4.8	3.3–3.5	108.9–113.7
$\bar{M}$	12.0	282.6	91.4	106.2	4.4	3.3	116.1
N	2	2	3	3	4	3	3
M₂ OR	—	—	78.0–83.5	97.0–130.9	3.6–4.0	3.0–4.2	116.1–152.1
$\bar{M}$	—	—	81.6	111.2	3.8	3.5	131.7
N	—	—	3	3	3	3	3
M₁ OR	8	176.6	53.0–60.0	92.1	4.3–6.2	2.1–3.0	174.0
$\bar{M}$	—	—	56.5	—	5.2	2.5	—
N	1	1	2	1	2	2	1

Palaeoloxodon recki, Osborn, 1942: p. 1275, fig. 1138.
Palaeoloxodon antiquus recki, MacInnes, 1942: p. 42, pl. 8, figs. 4, 5.
Elephas (Archidiskodon) recki, Arambourg, 1947: p. 252, pl. 1, fig. 4; pl. 2; pl. 3, figs. 1–4; pl. 4; pl. 5, figs. 1, 2, 5; pl. 6, figs. 1, 4; pl. 7, figs. 1–4, 6; pl. 8, fig. 5.
Elephas cf. *africanavus*, Leakey, 1965: p. 23, pls. 16, 17.
Mammuthus recki, Cooke and Corydon, 1970: p. 127, pl. 5, fig. B.

Lectotype: IPUB XVII 1384, left mandible with M_2. Figured in Dietrich, 1916: pl. 1, fig. 2; pl. 2, fig. 4.

Hypodigm: The lectotype and fifteen other syntypes, several hundred good molars, a number of skulls and abundant post-cranial remains. The following are outstanding specimens: KNM ER 352, right M_3; KNM ER 353, right M_3; KNM ER 342, left M_3; KNM (JK 2 A1116, Old. 4), mandible with right M_3; MNHN 33–9–331, right M^3; KNM (F3044), left M^3; KNM ER 341, mandible with left and right M_2; IPUB XVIII 1439–40, mandible with left and right M_2; KNM ER 351, left M^2; BM M-14695, right M^2; KNM ER 350, right M_1; MNHN (VJM ERUM 206), right M_1; KNM CH (068/6427), right M^1; BM M-14691, left M^1; MNHN 33–9–338, right dm_4; BM M-25/VI/35, right dm_4; KNM ER (1968/7), left dm^4; BM M-14942, left dm_3; TMP STS 1863, left dm_2–dm_3; KNM ER (1969 FS/94), left dm_3; IPUB GK 7.60, right dm^2; MNHM 1933–9–300, skull; MNHN 1933–9–775, partial skull and jaw.

Type of synonym: *Archidiskodon exoptatus*: IPUB Z.94–96, right M_3; lectotype chosen by the author (Maglio, 1969).

Horizon and localities: Later Pliocene to middle Pleistocene of Africa. The type is from Bed IV, Olduvai Gorge, Tanzania. Other specimens from Beds I–II, Olduvai; Kikagati, Uganda; Kaiso Formation, Uganda; Koobi Fora Formation, East Rudolf, Kenya; Shungura Formation, Omo Valley, Ethiopia; Ouadi Derdemi and Toungour, Lake Chad; Laetolil and Vogel River, southern Tanzania; Homa Mountain and Kanjera, Kavirondo Gulf, Kenya; Olorgesailie, central Kenya; Chemeron Beds, Baringo, Kenya.

Revised diagnosis: A medium-sized species of *Elephas* with molars of moderate width bearing 11 to 16 plates on M3; crown height generally 50 to 100 per cent greater than width, but may be as low as 20 per cent; anterior and posterior enamel loops strong in earlier evolutionary stages, becoming very irregular and weak in later stages; enamel 2.0 to 3.5 mm. thick on M3 and moderately to strongly folded; plates more closely spaced than in *E. ekorensis*, LF 3.9 to 6.6. Greatest width of molar about one-third up from crown base. Skull compressed in the facial plane; parietal and occipital regions greatly expanded; fronto-parietal surface more vertical in orientation than in *E. ekorensis*; occipital plane parallel to the facial axis; forehead concave; external naris very large, strongly downturned at the sides; strong fronto-parietal ridges terminating at widely spaced, but small orbits; premaxillary tusk sheaths massive, very closely spaced and parallel. Mandible short and massive, deep anteriorly; ramus broad; condyles rounded, condylar neck short. Plate formula:

$$M3 \frac{12-14}{11-16}, \quad M2 \frac{9-11}{9-10}, \quad M1 \frac{7-10}{8-9},$$

$$dm4 \frac{7}{8-10}, \quad dm3 \frac{?}{6}, \quad dm2 \frac{3}{3}.$$

Discussion: This is now one of the best known species of fossil elephant in Africa. Four evolutionary stages based on morphological changes and stratigraphic succession have been recognized (Maglio,

TABLE 13

Comparative Measurements for the Permanent Molars (M3-M1) of *Elephas recki* Dietrich 1916, in Millimeters. Specimens are Grouped into Four Stages as Explained in the Text.

	P	L	W	H	LF	ET	HI
Upper M³							
STAGE TWO							
KNM ER 342	12X[e]	233.2+	96.6[6]	119.5[9]	5.5	3.3 (2.9–3.6)	122.0
KNM ER 349	12X[e]	261.0[e]	91.6[6]	96.0[11]	4.7	2.7 (2.3–3.0)	126.0
MNHN 33-9-331	14X	265.0	88.7[6]	109.1[8]	5.0	3.4 (3.0–3.9)	123.8
MNHN 33-9-721	9+	178.0+	81.4[9]	127.0[9]	5.8	3.3 (3.0–3.6)	151.1
IPUB Vo. 330	3+	79.5+	77.0	119.0[e]	5.1	2.8 (2.3–3.4)	154.5
KNM [VM 307]	6+	109.4+	94.3[6]	113.0[6]	5.7	3.3 (3.0–3.6)	120.0
STAGE THREE							
KNM SHK II 244	4+	95.0+	76.0	124.0[e]	4.9	3.0	163.0
IPUB KL 2/39	9+	174.0+	83.2[2]	101.4[7]	5.0	2.8 (2.5–3.0)	140.0
KNM F3046	8+	157.0+	88.1[2]	141.3[5]	5.6	2.3 (2.2–2.5)	160.0
KNM SHK II 1294	7+	167.0+	97.0[2]	110.0[e]	4.9	3.0	120.0[e]
STAGE FOUR							
BM M-15418	14X	245.5	89.5[5]	147.5[11]	4.6	2.3 (2.1–2.4)	164.9
BM M-11977	13+	207.2+	82.0[1]	124.8[1]	5.8	2.8 (2.4–3.2)	152.2
KNM F3044	12X	242.0	81.6[7]	147.2[7]	5.3	2.0 (1.7–2.2)	181.0
BM M-14710	5+	97.0+	73.6	149.5	5.7	2.1 (2.0–2.2)	203.0
BM M-14713	3+	82.0+	105.7	197.2[e]	—	3.0 (2.9–3.0)	185.8
BM M-1407	+7+	129.2+	81.6	155.0	5.8	3.4 (3.1–3.7)	189.2
BM M-11977	12+	211.0+	87.0[3]	122.3[7]	5.8	2.6 (2.4–2.5)	140.8
Lower M₃							
STAGE ONE							
BM M-15212	8+	181.1+	85.1[3]	107.5[2]	4.7	3.0	126.6
BM M-15213	5+	121.0+	80.1[5]	104.8[e]	4.9	3.1 (2.9–3.3)	130.7
BM M-15211	12+	258.0+	84.5[6]	112.2	5.0	3.1 (2.8–3.4)	133.0
STAGE TWO							
KNM ER 352	12X	259.6	80.0[4]	100.2[6]	4.6	3.1 (2.7–3.4)	127.5
KNM ER 353	12X	241.0	93.0[8]	134.0[7]	4.6	3.2 (2.9–3.5)	144.0
KNM ER 342	15X	313.5	91.6[2]	109.5[11]	5.0	2.4 (2.0–2.4)	119.0
Omo coll. L 26-45	13	228.0	71.0[5]	85.0[10]	5.7	3.1 (2.9–3.4)	125.0+
MNHN 33-9-727	+4+	83+	78.6[2]	120.2[1]	5.0	3.3 (3.1–3.5)	153.0
IPUB Z94-96	12[e]	208.8+	86.6[3]	101.0[7]	5.1	3.3 (3.0–3.5)	142.3
Omo coll. L 26-49	12[e]	176.0+	59.5[4]	85.0[8]	5.9	3.0 (2.5–3.4)	142.0
STAGE THREE							
KNM HWK EII 874	9+	163.0+	75.0[3]	120.0[7]	5.9	2.4	160.0
IPUB XVIII 1439-40L	8+	213.6+	85.0[8]	136.5[4]	4.9	3.0 (2.6–3.4)	137.0
IPUB XVIII 1439-40R	7+	175.0+	71.3[7]	118.9[7]	5.0	3.0	166.6
KNM F3660 5/1	14[e]	298.0+	95.3[5]	140.5[8]	4.3	3.1 (2.8–3.3)	147.0
IPUB	9+	209.8+	92.4[6]	143.2[7]	5.3	3.1	155.4
STAGE FOUR							
KNM SHK 346	9+	176.0+	80.0[4]	111.0[8]	6.2	3.6	156.3
IPUB XVII 1384	13X	262.5	93.0[5]	133.8[4]	5.7	2.5	144.0
KNM JK II A1116	15X	334.5	87.7[6]	146.0[12]	5.0	3.3 (3.0–3.6)	170.0+
KNM JK II A1116	15X	339.0	88.0[5]	148.1[11]	5.4	3.0 (2.5–3.4)	170.0+
BM M-22663	4+	135.0+	81.0[2]	130.5[3]	—	3.2 (2.9–3.3)	161.2
BM M-11978	16[e]	315.0+	76.0[4]	123.2[4]	5.7	3.4 (3.2–3.5)	172.2
BM M-14699	8+	207.2+	77.1[4]	143.5[6]	5.2	2.5 (2.1–2.9)	186.4

1970a), but these grade into each other making formal taxonomic designation impossible. As discussed in an earlier section, the use of informal stage names proves as useful as subspecific recognition in terms of stratigraphic correlation, without a formal nomenclatural commitment. It is important, however, that the stage designation always accompany the specific name if this usefulness is to be realized.

The most primitive stage (stage 1) is known only from dentitions. The molar teeth show certain advances over the *E. ekorensis* grade; they are proportionately higher crowned, the plates are more numerous and more closely spaced, and the enamel is thinner (see table 13). This stage is known from Kikagati, Uganda, and probably from the lowest levels of the Shungura Formation (Coppens, pers. comm.).

Stage 2 is the typical form in the middle and upper beds of the Shungura Formation, above bed C. It also occurs at Koobi Fora and Ileret on the east side of Lake Rudolf and in the Laetolil and Vogel River deposits of the southern Serengeti. The plates are more closely spaced than in stage 1, the crown is proportionately higher, and the enamel is weakly folded, even in early phases of wear. Several skulls reveal important advances over the structure found in *E. ekorensis*.

The Olduvai Bed I and lower Bed II specimens of *E. recki* (stage 3) are slightly more progressive than stage 2. The crown is usually somewhat higher and the enamel more highly folded. The worn figures tend to be more irregular in outline with small, tight loops of enamel spaced without order around the plate.

The typical Bed IV Olduvai stage (stage 4), which also occurs at Olorgesailie and at Homa Mountain, is a higher crowned form with up to 16 plates on M3. The enamel is thinner and is usually intensely folded. The median enamel loops, when present at all, are very irregular in shape.

The identification to stage of any single specimen cannot usually be carried out with accuracy because these stages represent successional grades in a continuous evolutionary sequence, although the earliest stage can always be distinguished from the latest one. Adequate samples are required in order to minimize the effects of variability, which may be significant.

Elephas recki has been referred to at least five different genera from *Loxodonta* to *Mammuthus*, rarely with any reasons being given. From the generic diagnoses of *Elephas* and *Loxodonta* given above, and from that of *recki*, it is clear that this taxon is unrelated to the *Loxodonta* group. *Elephas* and *Mammuthus* are close in many cranial and dental characters, but they differ enough in skull morphology and tusk shape to rule out the latter as a generic assignment for *recki*.

Dietrich (1916) originally referred this taxon to *Elephas antiquus* ($=$ *E. namadicus* here) as a subspecies because of certain resemblances in molar structure. The most striking similarity is in the median enamel loop, but this is a primitive character for elephants generally, and not in itself indicative of close relationship. In *E. recki*, the molars are broader than in *E. namadicus* and are less evolved. The skulls of the two species are clearly related in structure, but are very distinct. In *E. namadicus* the skull is broader, the frontal bones are wider and shorter, the frontal crest is far more prominent and overhanging, and the premaxillaries are larger, with greatly flaring and well-separated tusk sockets. There can be little question as to the specific distinction between these two taxa.

The mandible described by Hopwood (1926) as *Elephas zulu* from near Kaiso Village is not separable from other specimens collected from the Kaiso Formation and I follow Cooke and Coryndon (1970) in including this material in *E. recki*.

Elephas iolensis Pomel, 1895
(Pl. X)

Elephas iolensis Pomel, 1895: p. 32, pl. 5, figs. 3, 4.
Archidiskodon transvaalensis Dart, 1927: p. 47, figs. 6 right, 7 left.
Archidiskodon sheppardi Dart, 1927: p. 48, figs. 6 left, 7 right.
Archidiskodon broomi Osborn, 1928: p. 672, fig. 2.
Archidiskodon hanekomi Dart, 1929: p. 713, fig. 24.
Palaeoloxodon kuhni Dart, 1929: p. 723, figs. 15, 16.
Pilgrimia wilmani Dart, 1929: p. 720, fig. 22.
Archidiskodon yorki Dart, 1929: p. 717, fig. 19.
Pilgrimia yorki Dart, 1929: p. 719, figs. 20, 21.
Pilgrimia archidiskodontoides Haughton, 1932: p. 4, pls. 1–2.
Pilgrimia subantiqua Haughton, 1932: p. 8, pl. 4, figs. 1–2.
Loxodonta (*Palaeoloxodon*) *darti* Cooke, 1939: p. 296, pls. 12, 13.
Elephas pomeli Arambourg, 1952 (in part): p. 413, fig. 6, pl. 1, fig. 3.

Holotype: Left M_2 in the Faculté des Sciences d'Alger, Algeria. (Currently in the Muséum national d'Histoire Naturelle, Paris.)

Hypodigm: The type and about two dozen teeth, mostly fragmentary. The best preserved material is as follows: MMK 3939, right M^3; MNHN (VJM EIUM 304), left M^3; SAM Mb 555, right M_3; MNHN (VJM EILM 205), left M_2; MMK 4074, right M_1; FSA (VJM 377), left M^1.

Types of synonyms: *E. transvaalensis*: right M^3, original destroyed. *E. sheppardi*: left M^3, original destroyed. *A. broomi*: MMK 3682, incomplete right M^3. *A. hanekomi*: MMK 3930, right M^3. *P. kuhni*: MMK 4144, left M_1. *P. wilmani*: MMK 4075, left M_2. *A. yorki*: MMK 4073, fragmentary ?M3. *P. yorki*: MMK 4074, right M_1. *P. archidiskodontoides*: MMK 4657, fragmentary right M^3. *P. subantiqua*: MMK 4286, left M_2. *L. darti*: partial M3, destroyed. *E. pomeli*: MNHN (VJM EIUM 304): left M^3.

Horizon and localities: Not entirely certain, but probably middle to late Pleistocene. The type is from Beausejour Farm, Algeria. Other specimens are from Port de Mostaganem, Algeria; Carrière Sidi Abder Rahmane (Casablanca), Morocco; Behanga I and Kaiso Village, Uganda (level of origin unknown); Natodomeri I and II, Kibish Formation, Sudan; younger gravels and terraces of the Vaal River, South Africa.

Revised diagnosis: A medium to large-sized species of *Elephas* with relatively broader molars than in *E. recki*; plates very thick, separated by thin cement intervals; crown height very great, generally 100 to 200 per cent greater than width; plate number low, 13 to 14 on M^3 (probably 16 to 17 on M_3); median loops usually absent on wear figures; enamel 2.0 to 3.5 mm. thick on M3, very strongly and irregularly folded. Plates

TABLE 14
Summary Measurements for the Permanent Molars (M3-M1) of *Elephas recki* Dietrich 1916, in Millimeters. Stages One and Two, and Stages Three and Four Are Grouped Together.

	P	L	W	H	LF	ET	HI
STAGES ONE AND TWO							
M³ OR	12–14	250.0–280.0	77.8– 98.1	99.5–153.0	4.5–6.3	2.7–3.4	117.0–163.0
M̄	12.5	263.7	90.6	121.7	5.1	3.0	135.3
SD	—	—	5.9	14.5	0.6	0.2	14.9
N	4	4	15	13	14	14	13
V	—	—	6.5	11.9	12.2	7.6	11.0
M₃ OR	11–15	228.0–313.5	59.5– 97.3	80.0–147.2	4.3–5.9	2.4–3.6	119.0–171.7
M̄	12.4	206.5	82.6	112.5	5.0	3.1	140.3
SD	1.2	—	8.5	17.1	0.6	0.3	15.6
N	8	4	25	18	24	24	17
V	9.6	—	10.3	15.2	11.2	10.3	11.1
M₂ OR	9–10	185.0–205.5	68.0– 79.1	97.0–109.8	5.4–5.9	2.3–2.9	140.0–148.0
M̄	9.3	198.5	74.0	103.7	5.6	2.6	144.0
SD	—	—	3.6	—	0.2	0.2	—
N	3	3	6	3	7	7	2
V	—	—	4.8	—	3.1	9.1	—
M¹ OR	7–8	131.0–143.0	62.1– 78.0	95.0	4.2–5.4	2.5–3.5	127.0
M̄	7.7	137.0	69.3	—	4.9	2.9	—
SD	—	—	6.6	—	0.6	0.5	—
N	3	2	5	1	6	6	1
V	—	—	9.6	—	11.7	17.0	—
STAGES THREE AND FOUR							
M³ OR	12–14	242.0–280.0	73.6–105.7	101.4–197.2	3.9–5.8	2.1–3.4	140.0–203.0
M̄	13.0	261.0	85.9	141.0	5.2	2.7	168.0
SD	—	—	9.2	25.8	0.6	0.4	21.2
N	2	2	11	10	11	11	10
V	—	—	10.7	18.3	11.8	15.3	12.6
M₃ OR	13–16	262.5–339.0	71.3– 97.8	111.0–148.1	4.3–6.2	2.4–3.6	137.0–186.4
M̄	14.6	312.0	86.3	134.7	5.1	3.0	186.4
SD	1.1	—	8.6	12.3	0.6	0.4	14.2
N	5	3	15	13	15	15	13
V	7.8	—	10.0	9.1	11.2	12.2	8.9
M₂ OR	9–11	172.1–221.8	70.0– 84.8	97.0–128.0	4.2–5.6	2.5–3.1	127.0–156.1
M̄	9.7	197.3	76.6	105.5	5.1	2.7	135.8
SD	—	—	6.0	—	0.6	0.2	—
N	4	4	6	4	6	6	4
V	—	—	7.9	—	12.2	8.1	—
M¹ OR	10	181.5	56.0– 86.1	95.0–108.5	5.5–6.6	2.2–3.0	126.1–189.0
M̄	—	—	69.5	102.5	6.2	2.5	152.7
N	1	1	3	3	3	3	3

closely spaced, lamellar frequency 3.4 to 6.3; greatest width at one-half up from base of crown. Faces of plates parallel, not widening significantly toward base.

Discussion: It is unfortunate that so little material of this species is available and that the skull is completely unknown. From the morphological and stratigraphic evidence, *E. iolensis* appears to have been the direct descendant of *E. recki* and for this reason transitional specimens may be impossible to assign to one species unit or the other. The majority of the available material referred to this taxon shows a marked advance in many traits over that of *E. recki*. The molars are broader and very much higher crowned, and the enamel is more highly folded. This species is the terminal member of the *E. ekorensis-E. recki* lineage.

Pomel's (1895) type specimen of *iolensis* was orig-

TABLE 15

Comparative Measurements for M3 of *Elephas iolensis* Pomel 1895, from North and South Africa, in Millimeters

	P	L	W	H	LF	ET	HI
Upper M³							
NORTH AFRICA							
MNHN	11+	204.7+	84.0⁴	170.1⁶	5.0	2.9(2.6–3.3)	201.3
MNHN	13X	245.0	98.9⁵	175.0¹⁰	6.0	3.0(2.7–3.2)	215.9
SOUTH AFRICA							
MMK 4157	13ᵉ	290.0ᵉ	113.0³	210.0¹⁰	5.0	3.5(3.0–4.0)	228.1
MMK 3930	14X	304.0	104.0⁵	218.0¹⁰	4.8	2.0	222.4
MMK 4523	5+	99.0+	96.0³	247.0⁴	4.5	2.0	257.0
MMK 4439	8+	130.5+	83.0³	186.0³	5.6	2.0(1.5–2.5)	224.0
Lower M₃							
NORTH AFRICA							
MNHM	11+	234.5+	73.9⁴	125.0+	4.4	2.5(2.1–2.9)	200.0+
SOUTH AFRICA							
SAM Mb555	9+	276.5+	96.0¹	177.0³	3.7	3.3(3.0–3.5)	186.4
MMK 10624	5+	163.0+	104.0⁴	206.0²	4.7	2.5	198.0
MMK 3682	6+	165.0+	114.0³	175.0⁶	3.4	3.8(3.5–4.0)	153.4

inally described as an M_3 of a dwarf species. Osborn (1942), accepting this identification, placed the species in the *E. namadicus* group, including the Mediterranean pygmy elephants. After examination of the type and other materials from North Africa, I have concluded that Pomel's identification was incorrect; the type is a second molar, not a third. This explains the small size and relatively low number of plates as compared with M_3's from South Africa referred to *Archidiskodon transvaalensis*. The structure of the broad, flattened posterior enamel ridge also tends to support this view. Several additional specimens from Port de Mostaganem and from Sidi Abder Rahmane show few significant differences from the South African material, especially when the high degree of variation in the two samples is taken into account (table 15). Although median expansions generally are lacking on M3 of *"transvaalensis,"* tiny ones are present on M1 and M2, as they are in the type specimen of *E. iolensis*. The South African sample eventually may prove to be distinct subspecifically, but there is no reason for specific separation from *E. iolensis*. The type specimens of *Archidiskodon transvaalensis* and *A. sheppardi* were destroyed, but several fine casts are available. Taken together, the southern material placed under these two groups show great variation in structure but there is little basis for maintaining the names as distinct. Thus, *A. sheppardi* should also be placed in the synonymy of *E. iolensis*.

Cooke (1960) placed the type of *Archidiskodon broomi* Osborn with other low crowned molars, primarily because of its thick plates and low hypsodonty index. However, the specimen is moderately well worn and the relative crown height was very much greater than preserved, probably HI was over 200. This species must be considered a synonym of *E. iolensis*.

Pilgrimia archidiskodontoides Haughton is known only from the very fragmentary type specimen (MMK 4657). As pointed out by Cooke (1960), the molar is unusual in having thick and very closely spaced plates. It is not possible to be certain of its relationships without additional material, but for the present it seems best to include the type within the present specific taxon.

Three South African species described by Dart (1929) as new are also referred to *E. iolensis*. The type and only known specimen of *Palaeoloxodon kuhni* and of *Pilgrimia yorki* were identified as third molars of small, thin plated forms. However, these are almost surely first molars and as such compare well with anterior teeth here assigned to *E. iolensis*. *Pilgrimia wilmani* is represented only by the type fragment, but is morphologically very close to *P. yorki* and appears to represent a left M_2 of *E. iolensis*.

In 1952 Arambourg (1952: p. 413) named as *Elephas pomeli* a specimen from Carrière Sidi Abder Rahmane which he distinguished from *E. iolensis* on the basis of its greater crown height. But, as already mentioned, the type of *E. iolensis* is an M_2, not an M_3. When compared with undoubted third molars of the present species, the type of *E. pomeli* cannot be separated on the specific level. Arambourg's referred specimens of *E. pomeli*, two lower second molar teeth from Oued Constantine, belong to *Loxodonta atlantica* as was discussed under that species.

A fragmentary molar described recently by me (*in:* Butzer *et al.*, MS.) was designated as *Elephas transvaalensis*. It was not possible in that short paper to discuss *transvaalensis-iolensis* synonymy in detail and the name commonly in use for teeth of this type in sub-Saharan Africa was employed. The importance of this material lies in the fact that it is the only positively

TABLE 16

Summary Measurements for Molars of *Elephas iolensis* Pomel 1895, in Millimeters

	P	L	W	H	LF	ET	HI
M³ OR	13–14	245.0–304.0	80.0–129.0	170.1–247.0	4.1–6.3	2.0–3.5	201.3–276.0
M̄	13.3	297.7	94.9	207.3	5.3	2.8	234.6
SD	—	—	15.9	26.0	0.6	0.6	17.5
N	3	3	9	8	9	9	8
V	—	—	16.8	12.6	11.3	21.4	7.4
M₃ OR	—	—	96.0–114.0	175.0–206.6	3.4–4.7	1.5–3.8	153.4–200.0
M̄	—	—	104.7	186.0	4.0	2.8	184.4
N	—	—	3	3	4	4	4
M₂ OR	12	215.9	63.2–92.5	133.6–174.0	4.0–5.9	1.0–2.5	162.0–217.7
M̄	—	—	74.4	147.7	5.2	2.0	194.4
SD	—	—	10.7	15.6	0.8	0.5	20.6
N	1	1	7	5	7	7	5
V	—	—	14.4	10.6	14.8	24.8	10.6
M₁ OR	—	—	—	—	—	—	298.6
M̄	8.0	180.5	79.0	—	4.8	2.3	—
N	2	2	2	—	2	2	1

dated specimen of *E. iolensis* known. The age of approximately 35,000 years B.P. (before present) indicates that the *Elephas* lineage survived in Africa until very late Pleistocene times. It is unfortunate that adequate stratigraphic data for other specimens are so largely lacking; it is unknown just how far back into the Pleistocene this species actually ranged. However, it seems unlikely that the earliest material presently known will prove to be much older than middle Pleistocene in age.

Elephas namadicus Falconer and Cautley, 1845
(Pl. XI)

Elephas namadicus Falconer and Cautley, 1845: pl. 13, figs. 1–3. 1847: pl. 12*A*; pl. 12*B*, figs. 1–3; pl. 12*C*, figs. 1–5; pl. 12*D*, figs. 1–3; 1846: 45.
Elephas antiquus Falconer and Cautley, 1845; pl. 13*A*, figs. 4, 5. 1846: pl. 13*B*, figs. 4, 5; pl. 14, figs. 1, 2. 1847: pl. 12*D*, figs. 4, 5; pl. 14*A*, figs. 1–13. (All mislabeled as *E. meridionalis* on original figures, but later corrected by Falconer (1868).)
Elephas priscus Falconer and Cautley, 1846: pl. 14, figs. 6, 7.
Elephas ausonius Verri, 1886: p. 453; Depéret and Mayet, 1923: p. 162, pl. 10, figs. 1, 2.
Elephas platyrhynchus Graells, 1897: p. 558, pl. 18, figs. 9*a* and 10.
Palaeoloxodon protomammonteus Matsumoto, 1924: 262, pl. 18, figs. 1, 2.
Loxodonta (Palaeoloxodon) tokunagai Matsumoto, 1924: p. 267.
Palaeoloxodon yokohamanus Tokunaga, 1934: p. 363, pl. 8, figs. 1, 2.

Lectotype: BM M-3092, skull with partial left and right M³ in place. Figured in Falconer and Cautley, 1847: pl. 12*A*.

Hypodigm: The lectotype and hundreds of additional specimens consisting of teeth, skulls, and post-cranial elements. May be found in almost all European museums. Good collections are housed in the British Museum (Natural History); Naturhistorisches Museum, Basel; Weimar Museum, Senkenberg Museum, Darmstadt Museum, Mainz Museum; Geological Survey of India, Calcutta; National Science Museum and the Geological Institute, Tokyo.

Types of synonyms: *E. antiquus:* left M_2 in mandibular fragment. Chosen by Osborn (1942: p. 1218) as lectotype. *E. ausonius:* IGF (no number), left and right M_3. *E. platyrhynchus:* maxillary fragment with partial molar and tusk, and skeletal elements. Present whereabouts unknown to author. *P. protomammonteus:* left M_3 in private collection of Mr. Natsume, Minato, Japan. *L. tokunagai:* MEO 2208, right M_3. *P. yokohamanus:* right M^2, whereabouts unknown to author.

Horizon and localities: Middle to late Pleistocene of Europe, Asia and the Middle East. The type locality is unknown, but is in the Narbadda Valley, central India. Other specimens are from the Godavari Valley, India; Mogok Caves, Burma; Choukoutien, Yenchingkou and Sjara-osso-Gol, China; Samarinda, Borneo; various localities in Japan; Middle Mosbach, Weimar, Taubach, Mauer and Steinheim, Germany; Gray's Thurrock and Upnor, England; Pignataro Interamna, San Romano, Pian dell'Ollmo and Malafrosca, Italy; numerous other localities in Europe and Asia.

Revised diagnosis: A large species of *Elephas* with narrow molars; 12 to 19 plates on M3; crown height generally 50 to 150 per cent greater than the width; worn enamel figures usually with pointed median expansions, but sometimes lacking; wear figures with central enamel portion and lateral rings in early stages of abrasion; enamel thickness 1.0 to 3.5 mm. with strong, close, but even folds; plates closely spaced with small cement intervals between them; lamellar frequency 4.4

TABLE 17

Comparative Summary Measurements for the Permanent Molars (M3-M1) of "*Elephas antiquus*" Falconer and Cautley 1946 and "*Elephas namadicus*" Falconer and Cautley 1846, in Millimeters

		P	L	W	H	LF	ET	HI
\multicolumn{9}{c}{"*Elephas antiquus*"}								
M³	OR	12–16	259.1–317.0	62.0–93.0	151.5–218.0	4.7–7.7	1.8–2.7	179.0–298.6
	$\bar{M}$	14.6	277.0	79.0	175.7	6.1	2.3	221.7
	SD	1.5	23.3	8.7	24.5	0.9	0.3	42.7
	N	8	8	10	9	10	10	9
	V	10.3	8.4	11.0	13.9	15.3	12.1	19.2
M₃	OR	13–18	232.6–339.0	50.0–88.0	123.5–166.0	4.4–6.8	1.8–3.4	145.0–302.0
	$\bar{M}$	15.4	289.1	72.3	146.9	5.7	2.6	208.8
	SD	1.8	39.7	10.1	12.3	0.7	0.5	49.5
	N	9	7	12	9	12	12	9
	V	11.7	13.7	13.9	8.3	12.8	20.5	23.7
M₂	OR	9–13	185.0–240.0	53.0–80.2	106.0–173.2	4.9–6.9	1.8–2.5	160.0–234.5
	$\bar{M}$	11.2	211.6	65.0	131.9	5.7	2.2	192.6
	SD	1.4	18.2	8.7	23.7	0.7	0.2	31.7
	N	10	8	11	6	11	11	6
	V	12.5	8.6	13.4	18.0	11.9	9.6	16.5
\multicolumn{9}{c}{"*Elephas namadicus*"}								
M³	OR	12–14	223.0	80.0–101.0	137.0–210.0	4.5–6.7	2.0–3.0	135.6–262.2
	$\bar{M}$	13.3	—	92.3	175.0	5.1	2.6	195.6
	SD	—	—	9.0	36.4	0.8	0.5	58.8
	N	3	1	6	5	6	5	5
	V	—	—	9.7	20.8	16.4	21.1	30.1
M₃	OR	14–15	284.0–333.0	79.0–86.0	132.0–144.0	4.5–6.7	3.0	153.4–177.4
	$\bar{M}$	14.3	308.5	83.5	138.7	5.2	3.0	166.1
	N	3	2	4	3	4	2	3
M₂	OR	9–14	242 0–292.0	66.0–99.0	119.0–126.0	4.3–6.8	1.6–3.0	180.4–188.4
	$\bar{M}$	12.0	255.4	77.7	122.5	5.3	2.4	184.4
	SD	2.0	21.2	11.1	—	0.8	0.6	—
	N	6	5	10	2	10	9	2
	V	16.7	8.3	14.3	—	15.9	23.6	—

to 7.7 for M3. Skull compressed in the facial plane but less so than in *E. recki;* occipital region greatly expanded posteriorly, flattened in the fore and aft direction; fronto-parietal surface expanded laterally and anteriorly with prominent overhanging frontal crest; forehead below crest very short and flat; temporal constriction very broad; external naris large; fronto-parietal ridge short; premaxillary tusk sheaths massive, flaring distally. Plate formula:

$$M3\,\frac{12-17}{13-18}, \quad M2\,\frac{11-13}{11-15}, \quad M1\,\frac{10-11}{9-11},$$

$$dm4\,\frac{10}{9-10}, \quad dm3\,\frac{6-7}{5-7}, \quad dm2\,\frac{?}{3}.$$

Discussion: The name *Elephas namadicus* first appeared on plate 13 (1845) of the volume of figures for the *Fauna Antiqua Sivalensis* of Falconer and Cautley (1845–1849); it was applied to a number of Asiatic specimens, including the lectotype skull. In plate 13*A* of the same 1845 publication, a mandible, presumably from a locality in Europe, was erroneously labeled *E. meridionalis* Nesti, but later (1868: p. 19) corrected by Falconer to *E. antiquus*. Since that time, the latter name generally has been applied to European material of this taxon, while *E. namadicus* has been used for specimens from Asia. Osborn (1942) maintained that these two geographic samples were unrelated because of supposed cranial differences seen in a skull from Pignataro Interamna, Italy, that lacked the prominent frontal crest of the Indian skull material. However, this specimen was erroneously reconstructed after excavation damage. New cranial evidence from Europe shows that this distinction can no longer be maintained.

In many ways the several known skulls from India (Pilgrim, 1905; Falconer and Cautley, 1847) are more different from each other than either of them is from European specimens. An undescribed skull from Pian dell'Olmo, Italy, in the Città Universitaria, Rome, is close to the type of *E. namadicus* and leaves little doubt as to the conspecificity of the European and Asiatic assemblages. There are no dental characters that would justify specific separation of the two samples, as can be seen from table 17.

Stratigraphically, the distribution of the Asiatic material has been assumed to be later than that from Europe. Present evidence on relative correlation between the two areas, however, indicates that the Narbadda Valley deposits and those of the Mogok Caves, Karewa Beds, and Yenchingkou are not significantly younger than the Cromer Forest Beds, Mosbach, or Süssenborn deposits (see section V below). Thus, there is no longer any reason to view the samples from Europe as specifically distinct from those of Asia. For this reason I have chosen to group both taxa under the species name *Elephas namadicus*.

A number of subspecies of *Elephas namadicus* have been described from Japan (Matsumoto, 1924, 1929; Tokunaga, 1934; Makiyama, 1924, among others). Although crania are lacking, the molar teeth seem specifically inseparable from *E. namadicus*, and the diagnostic characters, in most cases, can be ascribed to individual variation in width, height, and other parameters. The Japanese species *Palaeoloxodon protomammonteus, P. yokahamanus*, and *Loxodonta (Palaeoloxodon) tokunagai* are included in the present synonymy because the original diagnoses are insufficient to establish their distinctness from *E. namadicus*. However, the entire group of Japanese *Elephas* requires re-investigation.

Within Europe there is a great deal of variability among molars referred to this species. The stratigraphic evidence suggests that this species persisted through much of the middle and late Pleistocene, from the Cromer Forest Beds and Mauer to Taubach, a probable time span of nearly one million years. It is probably because of this that some specimens have been given specific recognition (e.g., as *E. platyrhynchus, E. ausonius*). However, when sufficient consideration is given to variation, these unusual specimens must be grouped with *E. namadicus* (see Osborn, 1942: pp. 1231–1232).

Hooijer (1961: p. 120) described several elephant teeth from Latemné in the Orontes Valley, Syria, as *E. trogontherii* (= *Mammuthus armeniacus* here). He held that the specimens could not belong to *E. namadicus* because they lack the median "loxodont" enamel figures and the lateral enamel rings of that species. His figure 3 does show prominent lateral rings on plates in early wear. Median expansions are highly variable in *E. namadicus* and their absence in the Latemné material cannot be considered unusual. The enamel is coarsely folded as in *namadicus*. This suggested occurrence of *E. namadicus* in the Middle East is significant for zoogeographic considerations, as we shall see in Section IX below. The elephant from Jisr Banot Yaqub, in Palestine, was also referred to *E. trogontherii* by Hooijer (1959). However, I refer it, tentatively, to *E. namadicus* for reasons similar to those discussed for the Latemné sample.

Elephas falconeri Busk, 1867 and other Mediterranean dwarfs

In a recent monograph on the insular species *Elephas falconeri*, Ambrosetti (1968) presents many morphological data on the dentition, skull and post-cranial skeleton of this species from Spinagallo, Sicily. He concludes that *E. falconeri* represents a valid species derived from an early stage of *E. namadicus*. Since I have not had the opportunity to examine this large collection, I defer to Ambrosetti's authority as it is reasonable on present evidence.

The validity and relationships of other described species of dwarf elephants from Mediterranean islands are less certain at this time. *Elephas mnaidriensis* Adams (1870) and *E. melitensis* Falconer (1862, 1868) from Malta, *E. lamarmorae* Major (1883) from Sardinia, *E. cypriotes* Bate (1903) from Cyprus, and *E. creticus* Bate (1907) from Crete seem to represent successive stages of dwarfing in species related to, and probably also derived from, *Elephas namadicus*. I see little biological reason why these species should be designated as distinct at this time, even though they occur for the most part on separate islands. The morphological differences between them need not reflect basic genetic divergence, being surely related to paedomorphic changes accompanying the different degrees of dwarfing.

Until a complete revision of all these insular forms is undertaken, their status will be open to question. Such a revision was not undertaken as part of the present work in view of current investigations by others. Nevertheless, whatever the ultimate taxomonic status of these forms, it seems almost certain that they are very closely related to each other and that the evolutionary trends from *E. namadicus* to *E. falconeri* discussed later would not be affected by recognition of structurally intermediate populations.

Elephas planifrons Falconer and Cautley, 1845 (1846)
(Pl. XII)

Elephas planifrons Falconer and Cautley, 1845: pl. 2, fig. 5; pl. 3, figs. 4–6; pl. 8, fig. 2; pls. 10, 11; pl. 12, figs. 1–13, pl. 13B, figs. 8–10.

Elephas hysudricus Falconer and Cautley (in part), 1845: pl. 8, figs. 2, 5.

Leith-Adamsia siwalikiensis Matsumoto, 1927: p. 213. Figured in Falconer and Cautley, 1845–1849: pl. 11, fig. 4; pl. 14, fig. 8.

Archidiskodon praeplanifrons von Koenigswald, 1951: p. 273, figs. 1–3.

TABLE 22
SUMMARY MEASUREMENTS FOR MOLARS OF *Elephas celebensis* HOOIJER 1949, IN MILLIMETERS. THIS IS THE PRESENT CONCEPT OF THE SPECIES AND INCLUDES SPECIMENS FORMERLY REFERRED TO "*Stegodon hypsilophus*"

	P	L	W	H	LF	ET	HI
M³ OR	8	156.0–139.6	42.0–64.0	49.0–57.0	6.0–6.5	3.7–3.9	87.0–116.0
$\bar{M}$	8.0	147.8	55.0	53.2	6.2	3.9	99.5
N	2	2	4	4	4	2	4
M₃ OR	11	164.0	41.0–44.0	46.0–47.0	7.5–8.0	2.5	109.0–115.0
$\bar{M}$	—	—	42.5	46.7	7.9	—	111.0
N	1	1	4	4	4	1	4
M² OR	9–10	120.0–145.0	42.0–47.0	36.0–45.0	7.5–8.0	2.8–3.5	—
$\bar{M}$	9.5	132.5	43.2	42.0	7.7	3.2	110.0
N	2	2	4	3	4	4	2
M₂ OR	7	—	56.6–57.0	36.4–38.1	5.8–6.2	2.5–3.1	—
$\bar{M}$	—	—	56.8	37.2	6.0	2.8	—
N	1	—	2	2	2	2	—

Pleistocene. The diagnosis reads like that of an elephant, and, in fact, the only supposed stegodont character was the compressed, Y-shaped valley between successive plates. The plane of section made to study the valley shape was taken in the midsagittal line of the molar. Examination of broken parts of these specimens reveals the presence of median columns that are not free at their apices except on the anterior one or two plates. A midsagittal section would have cut through these folds, giving a tightly compressed appearance to the valleys between adjacent columns. The *Stegodon*-like valleys in these specimens are, therefore, an artifact of sectioning. In all other characters, for example the apically sloping plate sides, the 3 to 6 apical digitations divided by prominent clefts, and the median columns, these molars are elephantine and not stegodont.

If we now compare these molars to other Asiatic species, the similarity to *E. celebensis* becomes obvious. They differ primarily in being slightly lower crowned, but this observation is based only on a few specimens and may not hold when a better sample is available. Even so, the difference is not great when the degree of variation of the known samples is allowed for. *Stegodon hypsilophus* must be considered a synonym of *Elephas celebensis*.

Elephas platycephalus Osborn, 1929

Elephas platycephalus Osborn, 1929: p. 22, fig. 21.
Platelephas platycephalus Osborn, 1936: p. 22; 1942: p. 1358, fig. 1219.

Holotype: AMNH 19818, skull with damaged right M³.

Horizon and locality: The age of the specimen is unknown, possibly early Pleistocene, more likely middle Pleistocene. The type and only known specimen is from the bed of the Amilee Creek, near Siswan, Simula Hills, India.

Diagnosis: (from Osborn, 1929: p. 22) Cranium of very primitive elephantine affinity, low, flattened; orbits widely separated from occiput; premaxillary rostrum somewhat broadened, resembling that of *Elephas*; posterior naris deeply indented; occipital condyles on relatively low plane, not greatly elevated above grinders; relatively long and narrow cranial proportions. Cranium widely different from the elevated *Elephas (Hypselephas) hysudricus* or the greatly elevated *Elephas indicus* crania. Ridge-plates of type molars fractured or absent. Estimated ridge-plate formula: M3 $\frac{16-\frac{1}{2}}{}$.

Discussion: The species is known only from the type skull of uncertain age. Morphologically, the specimen is distinct from all other elephants, but the significance of these differences cannot be determined without further material. On present evidence, we can only maintain *E. platycephalus* as a distinct species.

In 1942 Osborn diagnosed the new generic name, *Platelephas,* for this species. The characters given do not warrant more than specific rank, however. The cranium lacks the great elevation of the occipital region seen in many other species of *Elephas* and the tusk sheaths are directed downward and strongly forward. The orbits are large and far forward in position. The occipital plane slopes backward over the very low-placed condyles. The tusks were apparently parallel and of modest size, inclined only slightly to the facial axis. The palate is not depressed to any marked degree and the molars could not have been of very great height. The shape of the condyles and the laterally compressed and centrally constricted auditory bullae are typical of the genus *Elephas*.

In all of these characteristics, *E. platycephalus* most closely resembles the late Pliocene and early Pleistocene species *Elephas planifrons*. The relationships of this species remain in question, but it is possible that it represents a later stage in the *E. planifrons* specific lineage. If this is correct, its peculiar characters would be more readily understandable.

TABLE 23
SUMMARY MEASUREMENTS FOR MOLARS OF *Elephas hysudricus* FALCONER AND CAUTLEY 1845, IN MILLIMETERS

		P	L	W	H	LF	ET	HI
M^3	OR	12–15	235.0–302.0	93.0–107.0	108.0–137.2	3.9–6.5	2.5–4.8	112.5–147.2
	$\bar{M}$	13.5	267.7	97.2	125.4	5.4	3.2	131.1
	SD	—	—	5.1	11.4	1.2	0.9	14.9
	N	4	4	6	5	6	5	5
	V	—	—	5.3	9.1	22.0	28.6	11.4
M_3	OR	12–17	254.0–354.0	74.0–107.0	104.1–159.0	4.1–6.5	1.8–3.5	124.3–166.7
	$\bar{M}$	13.7	304.6	91.7	135.1	5.1	2.6	149.3
	SD	2.0	45.9	11.7	20.9	0.7	0.7	16.5
	N	6	5	12	5	12	11	5
	V	14.4	15.1	12.8	15.5	13.5	25.9	11.0
M^2	OR	9	—	92.0	—	4.5	2.3	—
	$\bar{M}$	—	—	—	—	—	—	—
	N	1	—	1	—	1	1	—
M_2	OR	—	190.0–211.0	71.8–76.0	93.1–97.3	4.9–5.3	2.7–3.1	128.1–135.4
	$\bar{M}$	10.0	200.5	73.5	95.2	5.1	2.9	131.7
	N	2	2	3	2	2	3	2
M^1	OR	—	151.1–180.0	78.0	107.6–111.2	5.5–7.2	2.2	142.7
	$\bar{M}$	9	165.2	—	109.4	6.4	—	—
	N	2	2	1	2	2	1	1
M_1	OR	9–10	121.0–180.2	52.0–78.0	81.8–111.2	5.1–8.2	1.0–2.7	129.6–42.7
	$\bar{M}$	9.3	150.7	66.5	100.0	6.3	2.1	136.1
	SD	—	—	—	—	1.3	0.7	—
	N	3	3	4	3	5	5	2
	V	—	—	—	—	20.6	30.8	—
dm^4	OR	8–10	108.0–129.5	48.9–70.6	70.9–97.2	6.3–10.0	1.1–2.2	137.7–145.0
	$\bar{M}$	8.5	116.7	58.1	83.4	8.0	1.5	142.4
	N	4	4	4	3	4	4	3
dm_4	OR	—	112.3–138.7	47.6–59.2	58.2–75.0	6.9–8.6	1.3–2.3	116.7–126.8
	$\bar{M}$	9.0	126.8	51.3	66.6	7.7	1.9	121.7
	N	3	3	4	2	4	4	2
dm^3	OR	—	—	41.0	—	9.8	1.0	—
	$\bar{M}$	—	—	—	—	—	—	—
	N	—	—	1	—	1	1	—
dm_3	OR	6–8	66.9–85.6	37.0–43.4	35.6–45.0	7.9–11.9	1.0–1.8	115.7
	$\bar{M}$	6.5	75.1	40.9	40.3	10.1	1.2	—
	SD	—	—	—	—	1.5	0.3	—
	N	4	4	3	2	5	5	1
	V	—	—	—	—	14.7	28.6	—
dm_2	OR	4–5	20.5–25.1	14.2–18.1	14.8–18.6	—	—	102.8–104.2
	$\bar{M}$	4.5	22.8	16.1	16.7	—	—	103.5
	N	2	2	2	2	—	—	2

Elephas hysudricus Falconer and Cautley, 1845
(Pl. XIII)

Elephas hysudricus Falconer and Cautley, 1845: pl. 1, fig. 3; pl. 4; pl. 5; pl. 6, fig. 1–3; pl. 7, figs. 1, 2, 5–12; pl. 8, figs. 1, 3–5; pl. 12B, fig. 4.

Elephas planifrons, Lydekker, 1886; p. 103, figured in Falconer and Cautley, 1845: pl. 8, figs. 2 and 2a.

Lectotype: BM M-3109, skull with third molars lacking crowns. Figured by Falconer and Cautley, 1845: pl. 4; pl. 5, figs. 1, 2. One of several cotypes designated by Lydekker: 1886; p. 117. Osborn (1942: pp. 1341–1342) referred to BM M-3127, an incomplete M^2, as the type because of page priority of the original figures. Although Lydekker's collective designation would technically be invalid under the rules of nomenclature if done today, I here take his first named specimen as the lectotype. Lydekker clearly established the concept of the species in designating these specimens

TABLE 24

Summary Measurements for Molars of *Elephas hysudrindicus* (Dubois) 1908, in Millimeters

	P	L	W	H	LF	ET	HI
M₃ OR	18	293.2–316.6	72.3–82.5	128.4–149.3	6.5	2.6–2.7	177.5–182.3
M̄	18.0	304.9	77.4	138.8	6.5	2.6	179.9
N	2	2	2	2	2	2	2
dm³ OR	7	56.9	31.6	42.7	7.1	1.4	135.2
M̄	—	—	—	—	—	—	—
N	1	1	1	1	1	1	1
dm₃ OR	7	65.5	30.9	34.0	12.1	1.2	111.0
M̄	—	—	—	—	—	—	—
N	1	1	1	1	1	1	1

as the types. It is also more desirable to have a skull, even one with incomplete teeth, as the lectotype than an incomplete, sectioned molar.

Hypodigm: The type and one hundred or more molar specimens and about ten skulls. The better collections are in the British Museum (Natural History), Indian Museum, Calcutta, and the American Museum of Natural History.

Horizon and localities: Latest Pliocene to lower Pleistocene of Asia. The type is from the Pinjor horizon, Upper Siwaliks, India. Other specimens are from the lower Karewa Beds, Kashmir, and the Upper Irrawaddy Beds, Burma.

Diagnosis: A medium to large-sized species of *Elephas* with relatively broad molars; 12 to 17 plates on M3; crown height generally 25 to 60 per cent greater than the width; plates thick and closely spaced, lamellar frequency 3.9 to 6.5 for M3; enamel moderate in thickness, generally 2 to 3 mm.; worn enamel figures irregular in outline; enamel weakly to coarsely folded in central portion of the figure; median loops weak to absent. Skull with very greatly expanded parietal and occipital regions; forehead transversely convex, vertically concave; deep median-sagittal depression; sharp parietal crests separating forehead from temporal fossae; frontoparietal ridges long, strongly directed downward and outward as in *E. maximus;* skull width at orbits less than occipital width; occipital plane inclined about 30 degrees to the facial axis; premaxillary tusk sheaths parallel, closely spaced and lying in the facial plane. Plate formula:

$$M3 \frac{12-15}{10-17}, \quad M2 \frac{9}{10}, \quad M1 \frac{9}{10},$$

$$dm4 \frac{8-10}{9}, \quad dm3 \frac{?}{6-8}, \quad dm2 \frac{?}{4-5}.$$

Discussion: Although unquestionably distinct, *E. hysudricus* has been confused with more progressive specimens of *E. planifrons* on molar evidence. Despite the overlap in molar morphology between the two species, they differ strikingly in cranial structure and must have diverged very soon after the genus entered Asia. By the time they first appear together in the Pinjor horizon, they are completely distinct. *E. hysudricus* has been questionably recorded from the Tatrot horizon (Kahn, pers. comm.); if this is confirmed, then the species would have overlapped completely with *E. planifrons*, both chronologically and spatially. The validity of the taxon has never been questioned and nothing further need be said here regarding its specific characters.

Elephas hysudrindicus Dubois, 1908

Elephas hysudrindicus Dubois, 1908: p. 1258. Hooijer, 1955b: p. 110, pl. 13, figs. 2, 3; pl. 14; pl. 15, figs. 1–5; pl. 16, figs. 1–3.
Euelephas namadicus, Martin, 1887: p. 53, pl. 6, fig. 3; 1888: p. 106, pl. 12, fig. 2.
Euelephas hysudricus, Martin, 1887: p. 57, pl. 6, fig. 2; 1888: p. 112, pl. 12, fig. 3.
Euelephas indicus, Dubois, 1891: p. 94.
Elephas ? maximus fossilis Van der Maarel, 1932: p. 168, pl. 16, figs. 1–3, text figs. 26, 27.

Holotype: DCR 4968–4969, skull and associated left third upper molar.

Hypodigm: The type and about 30 molar specimens of which the following are the best: DCR 2342, right M_3; DCR 2343, left M_3; DCR 11656, partial right M³; DCR 2352, partial right M³; DCR 2030, partial right M³.

Horizon and localities: Middle to late Pleistocene of Java. The type is from Tinggang, Solo Valley, Java. Other specimens are from Kedoeng Broeboes, Sangiran, Kedoeng Loemboe, Djambe, Bogo, and Ngandong, Java.

Diagnosis: A moderate-sized species of *Elephas* with narrow molars bearing 18 to 21 plates on M3; crown height 50 to 100 per cent greater than crown width; plates thin and closely spaced; lamellar frequency 6.5 for M3; enamel moderate in thickness, 2.5 to 3.0 mm. highly folded with small, open loops; wear figures irregular in outline; median loops and expansions lacking. Skull with broad, laterally, and anteriorly expanded parietal region, somewhat flattened dorsally as in *E. recki;* forehead concave in all directions, broader than long; temporal constriction broader than in *E. hysudricus;* parietal crests sharp; fronto-parietal ridges directed more outward than downward, shorter than in *E. hysudricus;* skull width at orbits greater than occipital width; external naris broad; premaxillary tusk sheaths parallel, well spaced as in *E. maximus.*

Discussion: This species was described by Dubois (1908) as a morphological intermediate between *E. hysudricus* and *E. maximus.* Although similar to the former in skull structure, there are clear differences between the two. The molar teeth are far more progressive in the Java species. Likewise, cranial morphology clearly distinguishes *E. hysudrindicus* from *E. maximus,* but on dental criteria, these two taxa are very close. Dubois suggested a phyletic relationship between *E. hysudricus, E. hysudrindicus,* and *E. maximus.* Until now this view has not been seriously challenged, but as we will see in section VI, *Elephas hysudrindicus* does not appear to have been ancestral to the living Asiatic elephant.

Elephas maximus Linnaeus, 1758

Elephas maximus Linnaeus, 1758: p. 11.
Elephas asiaticus Blumenbach, 1797: p. 124, no. 19, fig. C.
Elephas gigas Perry, 1811: pl. Ii.
Elephas sumatranus Temminck, 1847: p. 91; Osborn, 1942: 1329, fig. 1182.

Holotype: There is no type specimen.
Horizon and localities: Sub-Recent to Recent; known from many localities in Asia.
Diagnosis: A small to medium-sized species of *Elephas* with narrow molars bearing 22 to 27 plates on M3; crown height 50 to 150 per cent greater than width; plates thin and closely spaced, lamellar frequency 5.0 to 9.0; enamel moderately thick, 2.5 to 3.0 mm., coarsely folded with small, open loops; wear figures irregular in outline; median folds and expansions lacking. Skull with expanded parietal region; parietals rounded dorsally, not flattened; forehead weakly concave in all directions, longer than broad; temporal fossae bordered dorsally by sharp parietal crests; temporal constriction narrower than in *E. hysudrindicus;* fronto-parietal ridges directed more downward than outward; skull width at orbits less than occipital width; external naris narrower than in *E. hysudricus* or *E. hysudrindicus;* tusk sheaths small, parallel and well separated; tusks often lacking in females.
Discussion: This form, potentially the best known species of *Elephas,* is distinct in both cranial and dental morphology from every extinct form. It is hardly represented at all in the fossil record. The question of its ancestry is discussed below (see page 85).

Mammuthus Burnett, 1830

Mammuthus Burnett, 1830: p. 352 (for *E. primigenius*).
Dicyclotherium É. Geoffroy St.-Hilaire, 1837: p. 119 (for *E. primigenius*).
Cheirolites von Meyer, 1848: p. 286 (for *E. primigenius*).
Archidiskodon Pohlig, 1888: p. 138 (for *E. meridionalis*).
Parelephas Osborn, 1924: p. 4 (for *E. jeffersonii*).

Type species: *Mammuthus primigenius* (Blumenbach), 1799.
Included species: The type species; *M. subplanifrons* (Osborn), 1928; *M. africanavus* (Arambourg), 1954; *M. meridionalis* (Nesti), 1825; *M. armeniacus* (Falconer), 1857; *M. imperator* (Leidy), 1858, 1957; *M. columbi* (Falconer), 1857, 1922.
Distribution: Late Pliocene to early Pleistocene of Africa; early Pleistocene to sub-Recent of Europe and Asia; middle Pleistocene to sub-Recent of North America.
Revised diagnosis: Medium to large-sized elephants with broad molars in early forms, becoming narrower in some later species; 8 plates on M3 in earliest forms, up to 27 on latest species; worn enamel figures smooth to irregular in outline with strong median folds in earlier forms only; plates thick and widely spaced to thin and closely spaced, the lamellar frequency for M3 progressing from 3.0 in the earliest representatives to 11.0 in the latest; enamel thickness 5.5 to 1.0 mm., being very thin in the most progressive species; enamel smooth in early forms, becoming very weakly ribbed externally with minute, irregular wrinkling in later forms; crown height 75 to 300 per cent of the width, increasing in later species. Mandible in early forms with very retracted symphysis, short corpus, and low coronoid; mandibular tusks lacking; condyles rounded, set far back in early forms, progressively further forward in later ones. Skull very high with dorsally expanded parietals; foreshortened anteroposteriorly, greatly so in later species; midsagittal depression lacking; fronto-parietal surface concave vertically, convex transversely; temporal constriction strong; no parietal crests, temporal fossae bounded dorsally by gently rounded borders; fronto-parietal ridges long, directed downward and outward; orbits widely separated, skull width at orbits equal to or greater than occipital width; occipital plane inclined backward slightly in early forms, becoming more vertically to forwardly inclined in later forms; premaxillary sheaths narrowly spaced proximally, curving outward distally; tusks usually long, strongly curved, and spirally twisted.
Discussion: Despite the large number of specific and subspecific names that have been proposed for members of this group, diversity within it is not great and most of the valid species seem to represent successional stages along a single lineage. The specific distinctions used here are useful in identifying samples of specimens, but there is overlap in all characters. As more localities of intermediate age are found, we may expect specific distinction to dwindle even further, and definition of "species," if such units are then still desirable, will be even more arbitrary than it is now.

The skull and teeth of *Mammuthus,* particularly teeth of the earlier species, are morphologically close to those of *Elephas*. The important distinctions were discussed above. It was pointed out there that for these early species the skull and tusks are more useful in establishing relationships. Fortunately, skull and tusk fragments are now available for most of the earliest species of elephants and these allow a provisional reference to genus until better material becomes available. Thus, we can now trace the genus *Mammuthus* back into the Pliocene of Africa where it surely had a common ancestry with *Elephas*.

The molar teeth in *M. subplanifrons* and *M. africanavus* show similarities in structure to those of *E. planifrons;* all three taxa have thick, unfolded enamel and similar wear figures on the occlusal surface, all are low crowned, all have retained the primitive median enamel loops, all have a low number of plates, and all have similar proportions. But, in the later species of *Mammuthus,* the molar teeth are so distinctive that they are rarely mistaken for those of any other elephant group. The enamel is usually extremely thin and not thrown into tight folds as it is in *Elephas*. The crowns are very high, hence unlike the most progressive species of *Loxodonta*. In *M. primigenius,* the plates are thinner and more closely spaced than in any other elephant.

Although progressive changes occurred in the development of the skull, the peculiar twisted tusks so characteristic of the genus seem to have persisted from the earliest history of the group. Some variability is seen in tusk form, yet this character is very useful in distinguishing the genus from *Elephas* and *Loxodonta*.

In his monograph on the Proboscidea, Osborn (1942) separated the species of the present assemblage into three genera—*Archidiskodon* for the more primitive species *subplanifrons, meridionalis,* and *imperator; Parelephas* for intermediate species such as *columbi* and *armeniacus;* and *Mammonteus*[12] for the progressive *primigenius.* Osborn placed all three genera in his subfamily "Mammontinae" and distinguished between them by the relative degree of skull height, anteroposterior foreshortening of the skull, and palatal depth. *Parelephas* was intermediate in these structures between *Archidiskodon* and *Mammonteus*. Similar progressive distinctions were made on molar characters.

When the total degree of variability within this group is taken into consideration, we find essentially a continuous transition from *M. subplanifrons* to *M. primigenius.* There are some gaps in the record, but there is no basis for generic splitting. When a continuous lineage persists without basic adaptive shifts, generic divison seems only to obscure phyletic relationships. I, therefore, include both *subplanifrons* and *primigenius* in the same genus. Subgeneric distinctions may be found desirable at a later date, but I do not believe them to be useful at this time.

Mammuthus subplanifrons (Osborn), 1928
(Pl. XV, figs. 1–5).

Archidiskodon subplanifrons Osborn, 1928: p. 672, fig. 1.
Archidiskodon andrewsi Dart, 1929: p. 711, fig. 14.
Archidiskodon proplanifrons Osborn, 1934: p. 10, fig. 2.
Archidiskodon planifrons nyanzae MacInnes, 1942: p. 86, pl. 7, fig. 9; pl. 8, fig. 1.
Mammuthus (Archidiskodon) scotti Meiring, 1955: p. 189, pls. 1–4, text figs. 1–8.
Stegolophodon sp., Singer and Hooijer, 1958: pp. 1–3, figs. 1–4.

Holotype: MMK 3920, incomplete lower third molar, sectioned.

Hypodigm: The type and about two dozen molar specimens and several skeletal elements. The better specimens are as follows: MMK 4333, complete left M_3; NMB A.2882, left M_3, left ulna and left superior tusk (type specimen of *M. scotti*); BCUL (VJM MSUM 304), complete M^3; SAM L 12723, mandible with complete left M_3, fragmentary pelvis and atlas; BM KE 102, complete left M^3; BCUL (VJM 853), right dm^4.

Types of synonyms: *A. andrewsi:* MMK 435, probably M_1. *A. proplanifrons:* MMK 4333–4, left and right M_3. *M. scotti:* NMB A. 2882, left M_3, left tusk and left ulna.

Horizon and localities: Early Pliocene of eastern and southern Africa. The type specimen is from Sydney-on-Vaal, questionably from the middle terrace of the Vaal River, Transvaal, South Africa. Other specimens from the Vaal River terraces; Virginia, O.F.S.; Langebaanweg, Cape Province; South Nyabrogo, Kaiso Formation, Uganda; the Vogel River Series, Tanzania; locality J.M. 511, Chemeron Formation, Kenya; Kanam, Kavirondo Gulf, Kenya; and the Chiwondo Beds, Malawi.

Revised diagnosis: Molars very broad, with 7 to 9 plates on M3; crown height 60 to 90 per cent of width; plates moderately thick with wide intervening cement intervals; lamellar frequency 2.5 to 4.5 for permanent molars; enamel thick, 3.5 to 5.5 mm., not folded; worn enamel figures forming continuous loops with prominent median folds; crown divided into 4 to 6 apical digitations lacking prominent median clefts except on the first plate; posterior columns free at apices on anterior plates only. Mandible with long slender corpus; symphysis long; ramus high, but narrow. Plate formula:

$$M3\frac{7-8}{8-9}, \quad M2\frac{5-6}{?}, \quad dm4\frac{6}{?}.$$

Discussion: The early concept of this species was of a form related to, but more primitive than, *E. planifrons* of Asia. Osborn believed it to represent a species ancestral to all European archidiskodonts and to lie close to the very origin of the elephant phylum. Since that

---

[12] Osborn supposed this name to date from Camper, 1788. As Simpson (1945: pp. 134, 244) has pointed out, Camper's "*Mammonteus*" has no nomenclatural standing. *Mammonteus,* then, dates from Osborn, 1924, and cannot invalidate *Mammuthus* Burnett, 1830.

TABLE 25

COMPARATIVE MEASUREMENTS FOR LOWER M_3 OF *Mammuthus subplanifrons* (OSBORN) 1928, FROM SEVERAL LOCALITIES IN AFRICA, IN MILLIMETERS

	P	L	W	H	LF	ET	HI
VAAL RIVER							
MMK 3920,							
Type of *E. subplanifrons*	+5X	151.0+	102.0	80.0e	3.7	3.5(3.0–4.0)	78.5e
MMK 4334							
Type of *E. proplanifrons*	+5X	181.0+	104.0	75.0e	3.1	4.5(4.0–5.0)	72.7e
MMK 4333	8X	259.0	105.0⁵	70.0⁵	3.2	3.8(3.7–4.0)	77.1
VIRGINIA, O.F.S.							
NMB A2882							
Type of *A. scotti*	9e	196.0+	91.0⁵	73.0⁸	4.0	4.3(3.5–5.0)	89.0e
LANGEBAANWEG							
SAM L12723A	7+	264.6+	101.1³	66.6⁷	3.2	4.8(4.1–5.5)	66.8
SAM L12723B	9X	325.1	111.5⁷	76.3⁵	3.2	4.9(4.1–5.6)	68.4
KAISO							
BM M-26313	+3X	87.2+	88.5	—	—	5.4(4.8–6.0)	—
MARSABIT ROAD							
KNM	+3X	83.4+	90.2	64.0e	4.3	3.5(3.2–3.8)	71.0

time, the ancestral position of *subplanifrons* has been restricted to the mammoths only, and it is no longer considered to be the earliest member of the Elephantidae (Maglio, 1970a).

Too few specimens are available for an assessment of variability within the species. The high degree of variation appears to be due to differences between successive populations, since none of the known localities has yielded a large sample (table 25). These samples are distributed over a probable time span of nearly a million years and are far separated geographically. The Langebaanweg and Kanam East specimens generally fall within the lowest part of the observed range of known variability for the species and are in many ways transitional between *Primelephas* and *M. subplanifrons sensu* type specimen. The Kanam molars have only 7 plates plus a strong posterior enamel fold (which could be considered an 8th plate), the lamellar frequency is only 3 to 3.5, the enamel is thick (4 to 5 mm.), and the crown height is very low (HI = 65). In the Langebaanweg specimens the lamellar frequency is 3.2, the enamel thickness is 4.8 mm, and the hyp-

TABLE 26

SUMMARY MEASUREMENTS FOR MOLARS OF *Mammuthus subplanifrons*, IN MILLIMETERS

	P	L	W	H	LF	ET	HI
M^3 OR	7–8	202.8–229.2	81.0–122.0	49.0–85.0	2.6–4.1	3.4–5.0	56.3–91.0
$\overline{M}$	7.5	215.3	96.9	67.8	3.3	4.0	74.5
SD	—	—	14.7	14.2	0.6	0.6	16.2
N	4	3	7	7	7	5	6
V	—	—	15.2	21.0	16.7	15.9	22.8
M_3 OR	8–9	259.0–325.1	63.7–111.5	45.0–80.9	3.1–4.3	3.5–5.5	59.0–89.0
$\overline{M}$	8.5	292.0	94.4	67.4	3.5	4.5	72.7
SD	—	—	12.9	10.8	0.4	0.8	8.4
N	2	2	11	9	10	9	9
V	—	—	13.7	16.1	12.2	16.9	11.5
M^2 OR	5	146.5	92.0	—	3.5	4.4	—
$\overline{M}$	—	—	—	—	—	—	—
N	1	1	1	—	1	1	—
M_2 OR	6	85.0	68.0–68.2	43.0–47.0	4.6–4.8	5.0–5.5	66.0–67.0
$\overline{M}$	—	—	68.1	45.0	4.7	5.2	66.5
N	1	1	2	3	2	2	2
dm^4 OR	6	101.5	51.8	47.9	6.0	2.0	92.4
$\overline{M}$	—	—	—	—	—	—	—
N	1	1	1	1	1	1	1

sodonty index is 67 to 68 (Maglio and Hendey, 1970). In spite of the primitive nature of these specimens, they are within the expected range of variation of the taxon represented by the type specimen.

Meiring's (1955) type specimen of *M. scotti* (NMB A 2882) from Virginia, O.F.S., is distinctly more progressive than the type specimen of *subplanifrons,* having 9 plates on M3, a lamellar frequency of 4.0 and a relative crown height of 89. Other specimens referred to the species are transitional between the Langebaanweg and Virginia molars and all are, therefore, included in the same specific taxon.

In a recent discussion of the ancestry of the Elephantidae in Africa, it was shown that specimens formerly referred to *Stegolophodon* on that continent (e.g., by Singer and Hooijer, 1958) probably belong to the early stages of *M. subplanifrons* (Maglio and Hendey, *op. cit.*). In the same paper it was suggested that this species represented a heterogeneous assemblage with primitive molars whose relationships can only be clarified by discovery of cranial material. In their earliest stages, *Loxodonta, Elephas,* and *Mammuthus* undoubtedly possessed very similar molars. As all these groups were certainly present in Africa during the late Pliocene, it seems likely that molars of all three have been confused and grouped together. Nevertheless, the Virginia specimen, with its typically *Mammuthus* type of tusk, indicates that at least part of this assemblage represents an early stage of the mammoth lineage.

Mammuthus africanavus (Arambourg), 1952
(Pl. XV, fig. 6; pl. XVI)

Elephas africanavus Arambourg, 1952: p. 413, pl. 1, fig. 2; text fig. 1.
Elephas meridionalis, Pomel, 1895: p. 13, pl. 1, fig. 3.
Elephas planifrons, Depéret and Mayet, 1923: p. 120, pl. 4, fig. 7.
Loxodonta africanava, Coppens, 1965: p. 348, pls. 5–8; pl. 9, figs. 1–6.

Holotype: MNHN 1950–1–12, right M_3.

Hypodigm: The type, one fragmentary skull and less than 100 specimens of teeth. All specimens are in the Muséum national d'Histoire Naturelle, Paris.

Horizon and localities: Middle to late Pliocene of northern Africa. The type is from Lake Ichkeul, Tunisia. Other specimens are from Kebili, Tunisia; Fouarat and Oued Akrech, Morocco; Aïn Boucherit, Algeria; Toungour, Ouadi Derdemi and Koulá, Chad.

Revised diagnosis: A small species of *Mammuthus* with moderately broad molars having 9 to 12 plates on M3; crown height from 10 per cent less than the width to 20 or more per cent greater, rarely as much as 40 per cent; plates moderately thick and well spaced, the lamellar frequency being 3.6 to 5.6 for the permanent molars; enamel unfolded, 2.5 to 4.0 mm. in thickness; posterior enamel folds wearing as prominent median loops; sides of plates strongly tapering towards the crown apex. Skull of the *Mammuthus* type, with long, inwardly twisted tusks. Plate formula:

$$M3\frac{9}{10\text{--}13}, \quad M2\frac{8\text{--}9}{8\text{--}9}, \quad M1\frac{6\text{--}7}{7}, \quad dm4\frac{6}{6\text{--}7}, \quad dm3\frac{5}{?}.$$

Discussion: The molars of this species resemble those of the Siwalik species, *E. planifrons*. In both forms, the crown height is equal to the width, the lamellar frequency is about 3 to 5, and the enamel thickness is 3 to 5 mm. The number of plates on M3 is about 9 to 12 for both. Because of this similarity, *M. africanavus* has been considered a synonym of *E. planifrons* by Hooijer (1955), a proposal which seemed reasonable in view of the then-accepted occurrence of *E. planifrons* in Africa. This supposition was based on specimens now referred to *Loxodonta adaurora* and on several poorly known specimens (the names of which are now considered *nomina dubia*) from South Africa. However, with the large samples of molars now available, *M. africanavus* is distinguishable from the Asiatic species by the strongly tapering sides of its molar plates, and by the weaker median loops on worn enamel figures. A poorly preserved skull recently collected from Garet et Tir, Algeria, and referred to *M. africanavus* on molar evidence, bears a pair of typical *Mammuthus*-like tusks. From photographs of the specimen taken before it was damaged in transit from the field, the morphology of the skull is seen to be close to that of *M. meridionalis* (also see Arambourg, 1969–1970). The skull and tusks are strikingly different from those of *Elephas planifrons* and demonstrate without a doubt that these species were unrelated.

A large collection of elephant teeth from the Lake Chad area has been referred to *Loxodonta africanava* by Coppens (1965). The range of variation in this collection is very great, but the more complete specimens show a close affinity to *M. africanavus* (table 27). The HI is about 100 to 140, the lamellar frequency 3.5 to 5 and the enamel thickness is about 2.5 to 4 mm.; the enamel is not folded. There are about 12 plates on M3 and the worn enamel figures have prominent median loops. The general nature of these molars is more progressive than the Ichkeul and Kebili samples. Without skull material, it is not possible to determine with certainty whether these teeth belong to a very early form of *Elephas,* a progressive form of *Loxodonta adaurora,* or to a population of *M. africanavus*. Geographically, the Lake Chad region borders what is now the Sahara, but communication with the North African area during the early Pleistocene was probable. Additional evidence is needed, but until then, I follow Coppens in provisionally regarding this material as belonging to *M. africanavus*.

Mammuthus meridionalis (Nesti), 1825
(Pl. XVII)

Elephas meridionalis Nesti, 1825: p. 211, pl. 1, figs. 1, 2.
Elephas antiquus Falconer and Cautley (in part), 1845–1849: pl. 14B, figs. 1, 17, 18; pl. 42, fig. 19; pl. 44, fig. 19.

TABLE 27

SUMMARY MEASUREMENTS FOR PERMANENT MOLARS (M3-M1) OF *Mammuthus africanavus* (ARAMBOURG) 1954 FROM NORTH AFRICA AND FROM CHAD, IN MILLIMETERS

	P	L	W	H	LF	ET	HI
			NORTH AFRICA				
M³ OR	9	267.8	88.2–108.1	74.9–103.2	3.0–4.4	3.2–4.6	72.2–95.6
M̄	—	—	98.7	90.2	3.8	3.7	89.0
SD	—	—	9.0	—	0.5	0.5	—
N	1	1	5	4	5	5	4
V	—	—	9.1	—	13.7	14.3	—
M₃ OR	10–11	268.9–295.0	79.5–101.0	97.9–114.3	3.6–4.9	3.3–4.3	103.9–116.3
M̄	10.7	281.7	91.9	106.2	4.1	3.9	107.9
SD	—	—	8.7	5.9	0.4	0.3	5.2
N	4	3	8	5	8	8	5
V	—	—	9.5	5.6	9.5	7.7	4.8
M² OR	8	196.8–206.1	73.3–92.1	83.0–99.9	4.1–5.1	3.5–3.9	102.7–113.2
M̄	8.0	201.4	84.0	89.7	4.6	3.7	108.6
SD	—	—	7.4	—	0.4	0.2	—
N	2	2	5	4	5	5	4
V	—	—	8.8	—	9.5	4.9	—
M₂ OR	8	211.6–256.0	77.4–101.2	93.2–95.0	3.6–4.6	3.7–4.2	93.8–120.4
M̄	8.0	256.0	89.5	94.1	4.1	3.9	107.1
N	3	3	4	2	4	4	2
M¹ OR	6	138.3	62.5–70.0	66.2–70.0	4.8–5.5	2.3–3.0	100.5–109.1
M̄	6.0	—	66.2	68.2	5.0	2.6	105.0
N	2	1	4	3	4	3	3
M₁ OR	7	163.0	67.0–77.7	68.2–69.0	4.2–4.7	2.2–2.5	98.7–102.9
M̄	—	—	71.3	68.6	4.5	2.3	100.8
N	1	1	3	2	3	2	2
			CHAD				
M³ OR	—	—	81.8–95.8	98.5–110.0	3.9–4.0	3.3–3.5	114.5–119.4
M̄	—	—	88.8	104.2	3.9	3.4	116.9
N	—	—	2	2	2	2	2
M₃ OR	12	244.0	76.0–86.4	86.6–96.0	4.3–5.2	2.6–3.8	100.2–117.1
M̄	—	—	81.2	89.6	4.8	3.1	111.2
SD	—	—	4.5	—	0.3	—	—
N	1	1	5	4	5	3	4
V	—	—	5.6	—	7.1	—	—
M² OR	9	175.0	75.9–83.0	97.0–103.4	3.8–4.2	3.2–3.3	121.8–145.5
M̄	9.0	—	79.2	100.2	4.1	3.2	133.0
N	2	1	4	4	4	4	4
M₂ OR	9	—	65.8–93.1	83.9	3.4–3.8	3.3–3.4	127.2
M̄	9.0	—	78.3	—	3.6	3.3	—
N	2	—	1	1	2	3	1
M¹ OR	7	—	66.0–74.2	72.0–84.5	4.5–5.0	2.8–3.3	108.7–128.0
M̄	7.0	—	68.3	80.4	4.7	3.1	120.3
SD	—	—	—	—	0.2	0.2	—
N	2	—	4	4	5	5	4
V	—	—	—	—	4.1	8.0	—
M₁ OR	7	—	64.5–75.0	63.2	4.1–4.9	3.0–3.8	121.2
M̄	—	—	69.8	—	4.5	3.3	—
SD	—	—	—	—	0.3	0.3	—
N	1	4	4	1	5	5	1
V	—	—	—	—	7.1	10.1	—

TABLE 28

SUMMARY MEASUREMENTS FOR PERMANENT MOLARS (M3-M1) OF *Mammuthus africanavus* (ARAMBOURG) 1954, IN MILLIMETERS

	P	L	W	H	LF	ET	HI
M³ OR	9	267.8	81.8–108.0	74.9–110.9	3.0–4.4	3.2–4.6	72.2–119.4
M̄	—	—	95.9	94.9	3.9	3.6	98.3
SD	—	—	9.7	13.0	0.4	0.5	16.9
N	1	1	7	6	7	7	6
V	—	—	10.1	13.7	11.2	13.0	17.2
M₃ OR	10–12	244.0–295.0	76.0–101.0	86.6–114.3	3.6–5.2	2.6–4.3	100.2–117.1
M̄	11.0	272.2	87.8	98.8	4.3	3.6	109.3
SD	0.7	21.6	9.0	10.1	0.5	0.5	6.1
N	5	4	13	9	13	11	9
V	6.4	7.9	10.2	10.2	11.4	14.6	5.6
M² OR	8–9	175.0–206.1	73.3–92.1	83.0–103.4	3.8–5.1	3.2–3.9	102.7–145.5
M̄	8.5	192.6	81.9	95.0	4.4	3.5	120.8
SD	—	—	6.2	7.6	0.4	0.3	14.9
N	4	3	9	8	9	9	8
V	—	—	7.6	8.0	9.8	7.5	12.3
M₂ OR	8–9	211.6–256.0	65.8–101.2	83.9–95.0	3.4–4.6	3.3–4.2	93.8–127.2
M̄	8.4	226.6	84.7	90.7	3.9	3.7	113.8
SD	0.5	—	12.9	—	0.4	0.3	—
N	5	3	7	3	6	7	3
V	6.5	—	15.2	—	11.3	9.3	—
M¹ OR	6–7	138.3	62.5–74.2	66.2–84.5	4.5–5.5	2.3–3.3	100.5–128.0
M̄	6.5	—	67.3	75.2	4.9	2.9	113.8
SD	—	—	3.5	7.8	0.3	0.4	10.6
N	4	1	8	7	9	8	7
V	—	—	5.1	10.3	6.2	12.5	9.3
M₁ OR	7	163.0	64.5–77.7	63.2–69.0	4.1–4.9	2.2–3.8	98.7–121.1
M̄	7.0	—	70.5	66.8	4.5	3.0	107.6
SD	—	—	4.5	—	0.3	0.5	—
N	2	1	7	3	8	7	3
V	—	—	6.4	—	6.3	18.0	—
dm⁴ OR	—	—	50.8–62.8	43.2–50.1	5.4–6.9	2.2–2.3	69.9–98.0
M̄	—	—	58.5	47.7	6.2	2.2	82.6
N	—	—	3	3	4	3	3
dm₄ OR	6–7	87.9–104.1	49.1–59.2	36.2–40.0	6.1–7.9	2.0–3.0	75.9–78.1
M̄	6.2	96.5	53.3	38.1	6.9	2.3	77.0
SD	—	—	4.4	—	0.7	0.3	—
N	4	4	7	2	7	7	2
V	—	—	8.3	—	10.7	14.0	—
dm³ OR	5	65.1	40.1–42.8	37.5–37.8	6.6–8.9	1.7–2.3	89.8–93.5
M̄	—	—	41.7	37.6	7.4	1.9	91.6
N	1	1	3	2	3	3	2

Elephas lyrodon Weithofer, 1889: p. 79; 1890: p. 172, pl. 3, fig. 2; pl. 4, fig. 2; pl. 5, fig. 1; pl. 6, figs. 1–2.

Elephas planifrons, Depéret and Mayet, 1923: pp. 101–120, pl. 4, figs. 1–6, 8; pl. 5, figs. 1–5; text figs. 4–15.

Elephas planifrons rumanus Stefănescu, 1924: p. 179. Original figure (as *E.* cf. *meridionalis*) in Athanasiu, 1912 (1915): pl. 17, fig. 4.

Lectotype: IGF 1054, skull with left and right third molars. Figured in Nesti, 1825: pl. 1, figs. 1 and 2. Also in Osborn, 1942: figs. 858 and 861. Selected by Depéret and Mayet (1923: p. 126).

Hypodigm: Type and several hundred specimens consisting of skulls, jaws and teeth. Several skeletons are known from Europe and one from North America. The best collections are in the Geological Institute, University of Florence; Museum of Natural History, Basel; Muséum national d'Histoire Naturelle, Paris; and the British Museum (Natural History), London.

Types of synonyms: *E. planifrons rumanus:* left M₃ in Laboratory of Geology, University of Bucharest. *Elephas lyrodon:* a partial skull with tusks, and a second pair of tusks, in the Institute of Geology, University of Florence.

Horizon and localities: Middle Pliocene to early Pleistocene of Europe, middle Pleistocene of North America.[13] The type specimen is from the upper Val d'Arno, Italy. Other specimens are from the upper Val d'Arno; Montopoli and Laiatico in the lower Val d'Arno; Senèze, Chagny, and Durfort, France; Bacton and the Red Crag, England; Praetiglian and Tegelen Beds, Netherlands; Ferladani, Bessarabia, U.S.S.R.; many other localities in Europe; Aïn Hanech, Algeria.

Revised diagnosis: A species of *Mammuthus* with relatively broad molars bearing 11 to 14 plates on M3; crown height generally 10 to 60 per cent greater than the width; median enamel loops usually weak or absent, present on worn plates in some specimens; enamel moderately thick, 2 to 4 mm. for the permanent molars; plates thick and widely spaced; lamellar frequency 3.5 to 7.7. Skull with parietals and occipitals expanded dorsally; fronto-parietal surface strongly concave anteroposteriorly, flat to convex transversely; parietal crests lacking; external naris large, laterally elongated, slightly downturned at the sides; tusk sheaths closely spaced proximally, diverging distally and lying nearly in the frontal plane; tusks typically massive, strongly curved and twisted in adults. Plate formula:

$$M3 \frac{12-14}{10-14}, \quad M2 \frac{9-11}{8-10}, \quad M1 \frac{8-10}{9-10},$$

$$dm4 \frac{7-8}{8-9}, \quad dm3 \frac{5-6}{5-6}, \quad dm2 \frac{3-4}{3-4}.$$

Discussion: The existence in Europe of what appears to be an early stage of *M. meridionalis,* formerly confused with *Elephas planifrons,* is now widely accepted (Aguirre, 1969; Azzaroli, 1970; Maglio, 1970a). This stage seems to be confined to deposits of early Villafranchian age, occurring at Montopoli (Azzaroli, 1963) and Laiatico (Ramaccioni, 1936) in the lower Val d'Arno, and in the Praetiglian Beds of the Netherlands. Azzaroli (1970) suggests that this stage should be specifically distinct from the typical upper Val d'Arno elephant. However, so few specimens of it are known that adequate diagnosis is not possible at this time. The distinctive features of this early Villafranchian elephant, so far as known, are exactly those that set *M. africanavus* apart from middle Villafranchian *M. meridionalis* of the upper Val d'Arno. It seems to have been closely related to that species and could represent a European population of it. I prefer to retain it as an early stage of *M. meridionalis* in order to emphasize its clear relationship to the upper Val d'Arno material.

Typical *M. meridionalis* from the Montavarchi group (Azzaroli, 1970) of the upper Val d'Arno and numerous other localities in Europe is known from abundant cranial and skeletal remains and is clearly related to the mammoth species group. The twisted tusks and details of skull morphology are unmistakably *Mammuthus* in nature.

Specimens near the upper end of the range of variability for this species as defined approach and overlap with those of *M. armeniacus* (tables 29 to 31). When more localities of intermediate age become known it is likely that the two species will grade completely into each other on all morphological criteria. The elephant from Voigtstedt in Thüringia is intermediate between typical *M. meridionalis* and *M. armeniacus,* and was originally designated as *Archidiskodon meridionalis voigtstedtensis* by Dietrich (1958, 1965). This material could as easily have been considered an early stage of *M. armeniacus,* but for the present its retention in *M. meridionalis* is not unjustified.

The suggestion that these more progressive specimens of *M. meridionalis* should be given a new species name has been offered by Azzaroli (1970). If such a designation is formally proposed, it will, I believe, only lead to further confusion. For purposes of stratigraphic correlation, it would be better to designate evolutionary stage names for the successive forms of *meridionalis,* thus avoiding the proliferation of formal taxa that in themselves add nothing to our understanding of the species.

Accepting the view that three groups of *meridionalis*-like elephants succeeded each other in Europe during the early Pleistocene, I propose to designate three evolutionary stages for this species, as follows:

M. meridionalis—Laiatico Stage: Includes specimens from Montopoli and Laiatico in the lower Val d'Arno, and the Praetiglian Beds of the Netherlands. The specimens from Ferladani, Bessarabia, described by Pavlow (1910) as *E. planifrons* should be included here. There are only 11 to 12 plates on M3, the crown height is low with a HI of about 110, the lamellar frequency is 3.5 to 4, and the enamel is relatively thicker than in later stages (3.5 to 4 mm). Posterior enamel loops are usually strong and the enamel is only weakly folded, if at all.

M. meridionalis—Montavarchi Stage: Includes material from the upper Val d'Arno, Senèze, Chagny, and Grenada, among others. This is the typical form of the species as defined in the diagnosis given above. Posterior loops, although present on some molars, are often small or absent. The enamel is weakly to strongly folded in the median portion of the plate.

M. meridionalis—Bacton Stage: Includes material from the Bacton Forest Beds, Farnetta, l'Aquilla, Voigtstedt and probably Aïn Hanech, Algeria. This form has up to 14 plates on M3, a crown height of 140 to 165 per cent of the width, and relatively thin enamel. There are cranial differences as well (Azzaroli, pers. comm.).

---

[13] The North American mammoths and their probable synonymies are briefly discussed at the end of the present section.

TABLE 29

Comparative Measurements for M3 of the Three Stages of *Mammuthus meridionalis* (Nesti) 1825, in Millimeters. Specimens from Aïn Hanech, Algeria, Are Indicated with the Abbreviation (A.H.)

Upper M³	P	L	W	H	LF	ET	HI
LAIATICO STAGE							
IGF 1077	11X	267.2	109.1⁵	—	3.7	4.1(3.2–4.9)	—
IGF 1077	11X	260.6	108.6⁶	—	3.9	3.8(3.4–4.2)	—
MPP	+7X	163.2+	102.2	—	3.8	3.8(3.4–4.1)	—
MONTAVARCHI STAGE							
IGF 1054, type	12X	283.8	120.6⁶	—	4.4	3.3(2.8–3.7)	—
IGF 1054	13X	311.7	126.4⁵	—	4.2	3.6(3.2–3.9)	—
IGF 30	13X	228.8	92.0⁴	111.9⁹	5.7	3.1(2.7–3.5)	121.6
IGF 48	13	272.8	108.1⁵	136.2	5.9	2.9(2.4–3.4)	125.9
IGF 46	11X	254.6	94.4⁵	114.9⁸	5.5	3.5(3.1–3.4)	121.7
IGF Na5	12X	276.4	102.1⁴	112.8³	5.5	3.7(3.7–3.8)	110.5
IGF 12788	14X	286.3	93.2⁴	125.9⁸	5.1	3.4(2.9–3.8)	135.1
IGF 73	14X	266.8	125.0⁴	117.2⁸	4.8	3.1(2.8–3.4)	93.8
NMHN 1892-15	14X	315.0	125.5⁴	140.5¹¹	4.5	3.0(2.7–3.2)	111.9
BACTON STAGE							
BM M-10499	13X	284.6	116.5⁵	—	4.6	3.1(2.9–3.4)	—
AMNH 26998	14X	230.0	102.0⁴	—	6.1	—	—
MNHN 48-1-127 (A.H.)	13X	252.0	89.9³	127.6⁷	4.7	30(2.8–3.2)	141.9
MNHN 48-2-205 (A.H.)	12X	305.4	90.6²	133.2⁸	4.0	3.0(2.7–3.4)	147.0
MNHN 48-1-126 (A.H.)	13X	314.3	91.0⁵	139.0⁸	4.2	3.1(3.0–3.3)	152.7
MNHN 49-2-205 (A.H.)	12X	290.0	90.0²	128.0²	4.1	—	142.2

In his monograph on the North African Villafranchian, Arambourg (1969–1970) has raised to full species rank, as *Elephas moghrebiensis*, the Aïn Hanech material here referred to the Bacton Stage of *M. meridionalis*. This material generally is higher crowned and somewhat more progressive in plate number than typical *M. meridionalis* of the Montavarchi Stage, but they are well within the range of variability for all characters of the Bacton Stage of that species. Only cranial evidence will resolve this issue, but on present data, the Aïn Hanech material is best retained in *M. meridionalis*.

Mammuthus armeniacus (Falconer), 1857
(Pl. XVIII, fig. 1)

Elephas armeniacus Falconer, 1857: p. 319; 1863: p. 74, pl. 2, fig. 2.
Elephas trogontherii Pohlig, 1885; p. 1027; 1888: p. 193, fig. 79; p. 195, fig. 82.
Elephas intermedius Jourdan, 1891: p. 1013.
Elephas nestii Pohlig, 1891: p. 303; Osborn, 1942: p. 1059, fig. 941.
Elephas wüstii Pavlow, 1910: p. 6, pl. 1, figs. 1, 2.
Elephas antiquus trogontheroides Zuffardi, 1913: p. 130, pl. 9 (III), figs. 6a and 6b.

Holotype: BM 32250–52, two upper third molars and a portion of a lower molar.

Types of synonyms: *E. trogontherii*: right upper and lower M3 in the Schwabe Collection, Weimar. *E. intermedius*: no type was selected. The molars described by Jourdan have been lost and there are no figures of them. *E. nestii*: BM 39463, left M_3, and BM 27915, left M^3. *E. wüstii*: series of upper and lower molars (dm3 – M3) in the Cabinet géologique de l'Université de Moscou.

Horizon and localities: Middle Pleistocene of Europe. The type was found near Khanoos, Erzerum Province, Armenia, U.S.S.R. Other specimens are from Süssenborn, Mosbach, Mauer, upper Steinheim, Taubach, Ehringsdorf, Cromer Forest Beds, Swanscombe, Corton Sands, Clacton, Chelles, and many other localities.

Material: The type and many hundreds of specimens, of which good collections are in the Weimar Museum, Mainz Museum, Natural History Museum of Basel, and the British Museum (Natural History).

Revised diagnosis: Molars narrow, with 15 to 21 plates on M3; crown height varying from 40 to 200 per cent greater than width; plates thin, closely spaced with a lamellar frequency of 5 to 8 for M1 to M3; enamel figures lacking median loops, irregular in outline; enamel weakly ribbed externally, finely wrinkled, 1.5 to 3.0 mm. in thickness. Skull of *Mammuthus* type; parietals greatly elevated and skull anteroposteriorly compressed; occipital plane somewhat rounded behind, more vertical than in *M. meridionalis*, forming an acute angle with the frontal plane; premaxillary plane inclined downward to frontal axis with tusks directed downward. Plate formula:

$$M3\frac{14-21}{15-21}, \quad M2\frac{12-17}{11-14}, \quad M1\frac{11-13}{11-12},$$

$$dm4\frac{10-11}{10}, \quad dm3\frac{5-6}{7}.$$

TABLE 30

Summary Measurements for All Molars of *Mammuthus meridionalis* (Nesti) 1825, in Millimeters. These Data Include All Three Stages as Discussed in the Text

	P	L	W	H	LF	ET	HI
M³ OR	11–14	228.8–317.1	85.6–126.4	100.2–141.8	3.7–6.1	2.6–4.1	93.8–152.7
$\bar{M}$	12.8	273.0	104.8	122.7	4.9	3.3	125.6
SD	1.1	26.1	11.1	12.6	0.6	0.4	15.3
N	33	32	39	30	40	36	25
V	8.5	9.6	10.6	10.3	12.7	11.0	12.2
M₃ OR	10–14	212.0–306.0	69.1–119.4	75.0–152.0	3.5–5.9	2.4–4.1	107.8–165.5
$\bar{M}$	12.1	266.3	97.2	115.0	4.6	3.4	126.4
SD	1.5	27.5	12.4	21.0	0.6	0.4	15.4
N	19	18	36	17	36	33	14
V	12.3	10.3	12.8	18.3	12.8	12.0	12.1
M² OR	8–11	167.1–242.2	76.0–105.7	97.9–137.7	4.0–5.7	1.9–3.6	115.0–146.9
$\bar{M}$	9.7	206.7	86.8	113.9	5.1	2.9	130.6
SD	0.7	21.5	7.5	14.0	0.4	0.5	11.1
N	14	12	18	11	18	17	11
V	7.5	10.4	8.6	12.3	7.9	16.1	8.5
M₂ OR	8–10	190.0–209.0	69.0–97.0	90.9–111.2	4.6–6.1	2.4–3.5	104.6–141.4
$\bar{M}$	9.3	198.7	85.1	97.8	5.1	2.8	121.6
SD	0.8	9.6	7.9	8.0	0.4	0.4	16.7
N	6	3	10	5	10	8	5
V	8.7	4.8	9.2	8.1	8.2	13.9	13.8
M¹ OR	7–10	140.3–169.8	67.0–85.5	74.3–95.9	5.8–6.8	2.2–2.8	94.7–143.3
$\bar{M}$	8.7	154.0	72.3	87.0	6.1	6.1	123.7
SD	1.2	13.0	7.2	7.9	0.4	0.2	18.7
N	6	6	7	6	7	7	6
V	14.0	8.4	9.9	9.1	6.2	9.0	15.1
M₁ OR	8–10	116.8–185.0	56.5–73.4	93.1–94.9	6.6–7.7	1.9–2.6	129.4–130.8
$\bar{M}$	9.2	151.6	67.0	94.0	7.0	2.2	130.1
N	4	4	4	2	4	4	2
dm⁴ OR	7–8	107.0–156.5	46.0–75.0	—	6.0–7.5	1.5–3.0	—
$\bar{M}$	7.5	131.7	60.5	—	6.7	2.2	—
N	2	2	2	—	2	2	—
dm₄ OR	7–9	107.2–124.5	51.4–56.9	41.5–69.0	7.6–13.3	1.6–2.3	74.8–123.5
$\bar{M}$	8.2	116.6	54.4	53.0	8.9	1.9	101.3
SD	0.8	7.3	2.4	—	2.5	0.3	24.6
N	5	5	5	4	5	5	3
V	10.2	6.3	4.4	—	27.7	14.5	24.3
dm³ OR	5–6	52.4–63.0	32.2–38.1	31.7	9.7–10.0	1.3–1.4	98.4
$\bar{M}$	5.5	57.0	35.1	—	9.8	1.3	—
N	2	2	2	1	2	2	—
dm₃ OR	5–6	58.1–62.0	25.0–38.1	31.0–34.6	9.7–11.9	1.0–1.5	98.8–124.0
$\bar{M}$	5.3	60.0	33.4	32.8	10.9	1.3	111.4
N	3	3	4	2	3	3	2
dm² OR	3–4	21.2–22.2	17.9–19.9	14.6–19.4	—	—	81.6–97.5
$\bar{M}$	3.5	21.6	18.9	18.9	—	—	89.5
N	2	2	2	2	—	—	2
dm₂ OR	3–4	21.5–23.7	14.8–17.6	15.6–21.0	—	0.7	88.6–141.9
$\bar{M}$	3.5	22.6	16.2	18.3	—	—	115.2
N	2	2	2	2	—	—	2

Discussion: Clearly descended from *M. meridionalis*, this species is transitional, morphologically and stratigraphically, between the latter and *M. primigenius*. The earlier stages of *M. armeniacus*, such as Zuffardi's

TABLE 31
SUMMARY MEASUREMENTS FOR ALL MOLARS OF *Mammuthus armeniacus* (FALCONER) 1857, IN MILLIMETERS

		P	L	W	H	LF	ET	H/W
M^3	OR	14–21	213.0–358.0	57.0–107.5	118.0–218.0	5.0–8.2	1.5–3.0	145.3–304.9
	$\bar{M}$	18.6	293.0	85.2	162.5	6.5	2.2	199.7
	SD	1.9	33.9	14.2	28.0	0.8	0.4	40.3
	N	20	17	26	22	26	25	18
	V	10.0	11.6	16.7	17.2	12.7	16.9	20.2
M_3	OR	15–21	236.0–340.0	70.0–113.0	96.0–160.0	5.0–7.2	1.8–3.0	133.2–206.6
	$\bar{M}$	18.3	298.2	87.6	139.8	6.3	2.3	165.9
	SD	2.0	32.8	11.5	17.9	0.6	0.3	21.9
	N	15	14	21	21	22	20	16
	V	10.8	11.0	13.1	12.9	10.3	13.8	13.2
M^2	OR	11–17	168.0–240.0	55.0–95.0	105.0–162.0	5.5–7.8	1.3–2.8	123.2–285.5
	$\bar{M}$	13.3	206.7	77.4	134.3	6.9	1.8	186.9
	SD	1.7	23.5	12.7	20.8	0.8	0.8	49.1
	N	11	10	11	11	13	13	9
	V	12.6	11.4	16.4	15.5	11.2	26.9	26.3
M_2	OR	10–14	187.0–220.0	70.0–91.0	101.0–176.2	5.5–7.9	1.5–3.0	141.8–196.2
	$\bar{M}$	11.5	200.5	79.4	124.3	6.4	2.0	161.7
	SD	1.2	14.3	7.6	25.3	0.7	0.5	18.6
	N	8	8	8	7	8	8	6
	V	10.4	7.1	9.6	20.4	11.2	24.5	11.5
M^1	OR	10–12	131.0–169.0	54.0–71.0	70.0–112.0	7.4–9.5	1.0–2.0	162.3–162.7
	$\bar{M}$	11.0	144.5	64.2	87.7	8.4	1.3	162.5
	SD	—	—	7.3	—	0.8	0.4	—
	N	3	4	5	4	5	5	2
	V	—	—	11.4	—	9.9	34.4	—
M_1	OR	11–12	159.0–172.0	47.0–62.0	90.0–101.0	7.3–8.3	1.0–1.2	149.2–214.9
	$\bar{M}$	11.3	165.3	57.5	94.0	7.7	1.1	175.1
	N	3	3	4	3	4	4	3
dm^4	OR	10	—	45.0–47.0	55.0	10.4–10.9	—	117.0
	$\bar{M}$	10.0	—	46.0	—	10.6	1.0	—
	N	2	—	2	1	2	2	1
dm_4	OR	10	117.0–122.0	35.0–41.0	50.0–67.0	9.1–9.4	—	121.9–191.4
	$\bar{M}$	10.0	119.5	38.0	58.5	9.2	1.0	156.6
	N	2	2	2	2	2	2	2
dm_3	OR	5–6	60.0–64.0	23.0–32.0	30.0–37.0	9.9–11.0	1.0–1.9	120.0–160.8
	$\bar{M}$	5.7	62.0	27.7	33.5	10.5	1.2	140.4
	N	4	4	4	2	4	4	2

(1913) "*E. antiquus trogontheroides*" from near San Paolo de Villafranca, are very close to the latest stages of *M. meridionalis* and are easily confused with them. The latest occurrence of *M. meridionalis* seems to have been the Pastonian Crag at Bacton (Azzaroli, 1970); early stages of *M. armeniacus* are found in the Cromer Forest Bed and at Mosbach I.

Depéret and Mayet (1923) traced this species back through a series of morphological types to the San Paolo specimens of Zuffardi, pointing out its progressive characters over those of *M. meridionalis* and clearly distinguishing them from both *M. meridionalis* and *E. namadicus* (*op. cit.*, p. 179). They considered Jourdan's type of *E. intermedius* identical with Pohlig's *trogontherii* from Süssenborn, and confirmed its distinction from *M. primigenius*, a move which has not been contested seriously since then.

Pohlig's name *trogontherii* has been applied widely to this species in recent years. Aguirre (1969) recently pointed out that Falconer's type of *E. armeniacus* is certainly identical to it and has priority.

It is beyond the scope of this study to consider in detail the hundreds of referred specimens now available for this species. Thus, although its systematic position within the genus is not in question, intraspecific considerations must await further study; it may prove desirable at a later date to recognize distinct stages, as in *M. meridionalis*. The important point for the present is the transitional position of *armeniacus* between *meridionalis* and *primigenius,* which allows the examination of certain evolutionary trends within the genus, as discussed below.

Mammuthus primigenius (Blumenbach), 1803
(Pl. XVIII, figs. 2–4)

Elephas primigenius Blumenbach, 1803: p. 697.
Elephas mammonteus G. Cuvier, 1799: p. 21, pl. 5, fig. 2; pl. 6, fig. 1.
Elephas mammouth Link, 1807: p. 3845.
Elephas primaevus Blumenbach, *in* Adams, 1808: p. 152.
Elephas jubatus Schlotheim, 1820: p. 4.
Elephas paniscus Fischer de Waldheim, 1829a: pp. 285, 289.
Elephas periboletes Fischer de Waldheim, 1829a: pp. 285, 290, pl. 18, fig. 1.
Elephas pygmaeus Fischer de Waldheim, 1829a: pp. 285, 292, pl. 18, fig. 2.
Elephas campylotes Fischer de Waldheim, 1829a: pp. 285, 291.
Elephas kamenskii Fischer de Waldheim, 1829b: p. 276.
Mammut sibricum von Meyer, 1832: p. 64.
Elephas brachyramphus Brandt, 1832: p. xi. Figured by Tilesius, 1815: pl. 10.
Elephas giganteus Brandt, 1833. Figured by Breyne, 1741: pl. 1, figs. 1, 2.
Elephas commutatus Brandt, 1833. Figured by Cuvier, 1825: pl. 9, fig. 7 and p. 179.
Elephas stenotaechus Brandt, 1833: p. xiii.
Elephas platytaphrus Brandt, 1833: p. xiv. Figured by Cuvier, 1825: pl. 9, figs. 5 and 6.
Elephas macrorynchus Morren, 1834: p. 23, pl. 2, figs. 1–4.
Elephas odontotyrannus Eichwald, 1835: p. 723, pl. 63, figs. 1, 2.
Dicyclotherium primigenius É. Geoffroy Saint-Hilaire, 1837: p. 119, fig. 1.
Elephas americanus De Kay, 1842: p. 101, pl. 32, fig. 2.
Elephas kamensis de Blainville, 1845: p. 202.

Lectotype: Left lower M_3 in the Blumenbach collection of the Zoological Institute of the Museum of of the University of Göttingen; chosen by Osborn (1942: p. 1122, fig. 993 left). Osborn designated a milk molar from Osteröde, Germany, as "co-lectotype," but only the above-named specimen is here regarded as the lectotype.

Material: Thousands of specimens of teeth, many skulls and skeletons in most major museums in Europe, the Soviet Union, and North America.

Horizon and localities: Late Pleistocene of Europe, Asia, and North America. The lectotype is from Siberia, locality unknown. Other specimens are from numerous localities, among them Brundon, Suffolk, the lower and middle travertines of Ehringsdorf, the younger loess of Thiede in Brunswick, Emmendingen. (For North America, see page 63.)

Revised diagnosis: A relatively small to medium sized species of *Mammuthus* with 20 to 27 plates on M3; crown height 50 to 150 per cent greater than width; plates extremely thin and closely spaced, lamellar frequency 7 to 12 for M1 to M3; median enamel loops lacking; worn enamel figures irregular in outline, enamel very thin, 1 to 2 mm. in thickness and finely ribbed; crown heavily invested with cement; tiny anterior root, others coalesced. Skull extremely compressed anteroposteriorly, parietals very greatly elevated; occipital plane flat, meeting the parietals at a sharp, acute angle; forehead concave fore and aft, convex laterally, premaxillae strongly directed downward; tusk sheaths closely spaced. Plate formula:

$$M3 \frac{20-27}{20-27}, \quad M2 \frac{15-17}{15-16}, \quad M1 \frac{12-14}{11-15},$$

$$dm4 \frac{9-13}{10-11}, \quad dm3 \frac{8}{8}.$$

Discussion: The validity of this species was first brought to attention by Ludolf (1696: p. 92), who called it the Mammontovoi Kost. This determination has never been questioned. Although the earlier stages have been confused with *M. armeniacus,* Depéret and Mayet (1923) clearly distinguished *primigenius* from that species, which the revised diagnosis given above further demonstrates. Similarities between *M. armeniacus* and *M. primigenius* were pointed out by Falconer (1857). The direct phyletic relationship between them was established in 1881 by Leith-Adams, and most accurately stated by Soergel (1912). This phyletic sequence, for which the morphological and stratigraphic evidence is overwhelming, is attested to by the almost continuous gradation from one form to the other. The molar teeth in both are similar in basic structure, but are more progressive in *M. primigenius.* In the skull, *M. primigenius* carries to an extreme those adaptations initiated in *africanavus,* further developed in *meridionalis,* and already well advanced in *armeniacus* (for further discussion, see pages 100–101).

The confusion in the early literature on this species requires some clarification so as to avoid possible future complication. In 1799 Blumenbach published, without figures, a description of an elephant from Burg-Tonna, Gotha, Germany, as an example of his new elephant from Siberia and Germany, but he did not apply a specific name at that time. In the same year, Cuvier (1799) applied the name *Elephas mammonteus* to the Siberian mammoth. Blumenbach later used the name *E. primigenius* in a French translation of his 1799 description (1803: 2: p. 407), yet the only material distinctly specified remained the Burg-Tonna skeleton. Cuvier (1806) later abandoned his name *mammonteus* in favor of Blumenbach's designation. According to Osborn (1942: p. 1122) the Burg-Tonna specimen belongs to *Elephas namadicus* which would give Blumenbach's name priority for that taxon if the Burg-Tonna specimen were taken as the type. Cuvier's *mammonteus* would then become the valid name for the Siberian mammoth. In 1942, Osborn (1942: p. 1122) selected an incomplete molar from Siberia as the lectotype of *M. primigenius,* thus fulfilling *his* concept of that name, but this specimen is specifically distinct from the skeleton selected by Blumenbach as *typical* for the species.

Application of strict priority cannot be used here

TABLE 32
Summary Measurements for All Molars of *Mammuthus primigenius* (Blumenbach) 1897, in Millimeters

	P	L	W	H	LF	ET	HI
M³ OR	20–27	226.0–285.0	68.0–113.0	135.0–188.5	6.5–11.1	1.3–2.0	164.6–211.8
$\overline{M}$	23.2	263.8	92.5	168.1	9.0	1.6	184.4
SD	1.9	20.5	10.8	16.4	1.3	0.3	14.3
N	12	7	17	15	17	17	15
V	8.2	7.8	11.7	9.7	14.5	16.3	7.8
M₃ OR	20–25	207.0–320.2	65.0–100.0	123.0–184.1	6.8–10.2	1.3–2.0	137.8–189.2
$\overline{M}$	21.8	267.4	87.6	137.8	8.5	1.5	159.7
SD	1.9	44.1	10.9	20.9	1.1	0.3	19.8
N	5	5	8	8	8	8	8
V	8.8	16.5	12.5	15.2	13.2	18.1	12.4
M² OR	15–16	154.0–172.0	64.0–80.0	127.0–151.0	9.4–11.5	1.0–1.3	198.4–228.8
$\overline{M}$	15.5	163.5	70.7	141.2	10.3	1.3	214.7
SD	0.5	6.3	6.8	—	0.9	0.1	—
N	8	6	7	4	8	7	4
V	3.4	3.9	9.6	—	8.4	9.0	—
M₂ OR	15–16	147.0–185.0	43.0–85.0	100.0–136.0	7.6–11.4	1.0–2.0	159.8–232.6
$\overline{M}$	15.3	174.3	67.0	121.9	9.2	1.3	197.0
SD	0.5	15.5	12.8	11.4	1.2	0.3	29.8
N	7	5	10	7	9	8	6
V	3.2	8.9	19.1	9.3	13.5	26.0	15.1
M¹ OR	12–14	122.0–154.5	48.0–76.0	99.0–123.5	10.3–11.0	1.0–1.4	206.3–208.3
$\overline{M}$	12.7	138.2	57.3	107.5	10.7	1.1	207.3
N	3	2	3	3	3	3	2
M₁ OR	12–15	124.0–146.0	41.0–76.9	69.0–104.0	7.7–11.0	1.0–1.7	168.3–192.5
$\overline{M}$	13.3	135.0	56.2	91.7	9.9	1.2	180.4
N	3	2	4	3	4	4	2
dm⁴ OR	10–13	102.0–121.0	37.0–57.9	60.0–84.0	8.2–16.0	1.0–1.5	141.3–189.2
$\overline{M}$	10.4	109.0	46.6	69.7	11.7	1.1	189.2
SD	1.1	9.0	7.3	10.3	2.3	0.2	26.7
N	7	6	6	4	7	7	3
V	10.9	8.2	15.7	14.8	19.4	17.4	16.8
dm₄ OR	10–11	98.0–105.0	37.0	65.0–70.0	11.4–11.5	1.0	189.2
$\overline{M}$	10.5	101.5	—	67.5	11.4	1.0	—
N	2	2	1	2	2	1	1

without causing great confusion. In spite of the fact that Blumenbach's concept of his new species was the Burg-Tonna specimen, his original name has been used for the woolly mammoth almost since its introduction in 1803, and it should be retained.

NORTH AMERICAN MAMMOTHS

The elephants of North America have been greatly confused in the past. Numerous generic and specific names have been applied to them and diagnoses of specific taxa have usually been inadequate. Even those names in most common use today have not been sufficiently defined nor have they been stable in their usage.

I have not attempted a detailed study of North American collections for the present review. The large and excellent collections now available require a comprehensive analysis which could form the basis of a monograph in itself. For the purpose of completeness, however, I shall briefly discuss the North American species in relation to the Old World elephant faunas as examined above.

The mammoth affinities of the New World elephants have long been recognized. Osborn (1936, 1942) recognized 16 specific taxa of Pleistocene mammoths in America which he grouped into three distinct genera of complex phyletic relationship. Of these species only 3 warrant species level recognition here. All available evidence suggests that these taxa represent a series of more or less successional populations in which progressive evolutionary change in masticatory adaptations paralleled those of the European mammoth lineage. On present taxonomic usage, all North American mammoths must be included in the genus *Mammuthus*. It is important to keep in mind that "species" which represent segments along a phyletic continuum are arbitrary units. The more complete the record of transitional populations, the more arbitrary become the species limits. Thus, as was the case for the European mammoths, the separation of distinct taxa of North American elephant taxa is less than satisfactory and the identification of an individual specimen is not always clear-cut. It is more important to be able to place any sample within the evolutionary sequence than to be able to attach a formal name to it.

A number of primitive-looking specimens from Pleistocene deposits of Idaho, Kansas, and Nebraska are the earliest elephants known from North America. Morphologically, these specimens are close to the Bacton Stage of *M. meridionalis* from nearly contemporaneous deposits of Europe. The earliest dated record in the New World is from Bruneau, Idaho, where interbedded basalts have been dated at 1.36 m.y. (Malde and Powers, 1962). The most complete specimen known is a partial skeleton found near Angus, Nebraska, and described by Osborn (1932) as *Archidiskodon meridionalis nebrascensis*. The Angus elephant is probably Irvingtonian in age, but the lack of associated fauna with this specimen makes any age determination uncertain.

Other localities have yielded specimens not differing greatly from the Angus skeleton. These have been described under several names, including "*Archidiskodon haroldcooki*" Hay (1928) from Holloman's gravel quarry, Oklahoma, and from the Arkalon local fauna of Kansas, and "*Archidiskodon hayi*" Barbour (1915) from Crete, Nebraska. Although this mammoth is poorly known at present, all available specimens seem to fall within the range of variability of a single specific taxon. Variation grades upward into more progressive forms. The robust structure of the type mandible of "*A. haroldcooki*," in particular, forecasts the massive jaw of *M. imperator* of a somewhat later age. The molar teeth are rather low crowned, with 10–13 plates on M3. The enamel is thicker than in any other American elephant (2.5–3.0 mm.) The lamellar frequency is only 4.0–4.5. In all of these features, this taxon cannot be separated from *M. meridionalis*.

In 1858 Leidy described a new species, *E. imperator*, based on a poorly preserved and very fragmentary specimen from the Niobrara River, Nebraska. The holotype is insufficient to establish an adequate diagnosis. In 1922 Osborn selected an incomplete neotype specimen (AMNH 11871) from Guadalajara, Jalisco, Mexico, which clearly established his concept of that species. The neotype compares well with Leidy's type as far as any comparison can be made. A large number of specimens from the southern United States and Mexico are now referred to this taxon. The molar teeth closely resemble those of the *M. armeniacus* stage of Europe, and although both of these taxa were certainly derived from *M. meridionalis*, this North American form is usually given specific recognition and lies at the base of a distinct phyletic branch. Indeed, its broad molars and massive mandible would seem to support this view. The species is first recorded in middle Irvingtonian deposits of the western United States. Successive stages show progressive increase in molar plate number over the condition in *M. meridionalis*, with 16–19 plates on M3 in the typical form. Later occurrences grade morphologically into the more progressive elephant of the late Pleistocene, *M. columbi*.

The great confusion associated with the name *M. columbi* resulted in part from Falconer's inadequate holotype specimen and from Osborn's (1922) selection of two neotype specimens (AMNH 13707) both of which are very close to *M. imperator*, if not actually identical to it. Osborn concluded that the holotypes of *imperator* and *columbi* were probably conspecific, although in later publications he retained both names. For the more progressive elephant material that had previously been referred to *M. columbi*, Osborn proposed the specific name *jeffersonii*. Although Osborn was correct in considering Falconer's original holotype specimen as inadequate for species diagnosis, there is little evidence that his neotype accurately reflects the true characters of the original. Thus, it is probably best at present to retain Leidy's name *imperator* for the more primitive of these mammoths and Falconer's name *columbi* for the more progressive stages. This also conforms with the most common usage of these names.

The major trends along the phyletic lineage from *M. columbi* to *M. jeffersonii* were an increase in the number of molar plates, an increase in the lamellar frequency, reflecting closer packing of the plates, and a concomitant thinning of enamel.

Mammuthus jeffersonii is the typical elephant of the Sangamon and Wisconsin. Successive stages exhibit progressive evolution in the shearing specializations of the dentition and in the configuration of masticatory muscles. The third molar teeth of the typical southern populations have 20–24 plates, a lamellar frequency of 5–7, and an enamel thickness of 2.0–3.0 mm. From this form a more southerly distributed race apparently evolved in which dental specializations were carried to an extreme. The number of plates was 24–30 for M3, the lamellar frequency was 7 to 9 and the enamel thickness was reduced to 1.5–2.0 mm. So closely did these features parallel those of *M. primigenius*, that the two taxa are often impossible to separate on dental characters alone. This very progressive stage of *jeffersonii* cannot be separated specifically from earlier stages which grade into it. When studied in more detail it may prove better to view the species as polytypic, with the northern race modified for feeding on boreal and tundra vegetation.

The following taxa, then, are provisionally suggested

as valid for North America (synonymy given only for North American specimens):

Mammuthus meridionalis (Nesti), 1825

Elephas meridionalis Nesti, 1825: p. 211, pl. 1, fig. 1, 2.
Archidiskodon hayi Barbour, 1915: p. 129, figs. 1, 3.5d.
Archidiskodon haroldcooki Hay, 1928: p. 33; Hay and Cook, 1930: pl. 3, fig. 1; pls. 13, 14.
Archidiskodon meridionalis nebrascensis Osborn, 1932: pp. 1–3, figs. 1–3.

Lectotype: IGF 1054, skull with left and right third molars (see page 55).

Types of synonyms: *A. hayi:* USNM 23-6-14, mandible with left and right M3 in place. *A. haroldcooki:* CMD 1057, mandible with left and right M3 in place. *A. meridionalis nebrascensis:* CMD 1359, nearly complete skeleton lacking cranium.

Horizon and localities: Early and middle Pleistocene of North America and Europe. The type is from the upper Val d'Arno, Montavarchi group, Italy. North American specimens are from Bruneau, Idaho, Angus, Nebraska, Crete, Kansas, Holloman's Quarry, Oklahoma, and several other localities.

Mammuthus imperator (Leidy), 1858

Elephas imperator Leidy, 1858: p. 2; 1869, pl. 25, fig. 3.
Elephas exilis Stock and Furlong, 1928: p. 140; Stock, 1935: p. 210; fig. 6.
Archidiskodon sonorensis Osborn, 1929: p. 18; fig. 18.

Holotype: USNM 185, fragmentary M^3.

Types of synonyms: *E. exilis:* CIT 14, partial skull and jaw, with both left and right M3 in place. *A. sonorensis:* AMNH 22637, palate with left and right M_3 in place, and partial jaw with right M_3.

Horizon and localities: Middle Pleistocene of North and Central America. The type locality is from the Loup Fork, Niobrara River, Nebraska. Other specimens from numerous localities in southeastern and western United States, Mexico and southward to Panama.

Mammuthus columbi (Falconer), 1857

Elephas columbi Falconer, 1857: p. 319; 1863: p. 43, pl. 1.
Elephas jeffersonii, Osborn, 1922: p. 11, fig. 10.
Elephas roosevelti Hay, 1922: p. 100. Figured in Osborn, 1942: fig. 968.
Elephas washingtonii Osborn, 1923: p. 4; 1942: p. 1101, figs. 972, 975, 893B and B1.
Parelephas progressus Osborn, 1924: p. 4; 1922, figs. 11, 12.
Elephas eellsi Hay, 1926: p. 154, figs. 1, 2.
Elephas floridanus Osborn, 1929: p. 20, fig. 20.

Holotype: BM 40769, fragmentary right M^3.

Types of synonyms: *E. jeffersonii:* AMNH 9950, complete skeleton. *E. roosevelti:* USNM 2195, complete upper and lower M3. *E. washingtonii:* AMNH 8681A, mandible with left and right M_3 in place. *E. progressus:* AMNH 10457 (Warren Collection), upper and lower, left and right M3. *E. eellsi:* fragment of a skull and tusks, present whereabouts unknown to author. *E. floridanus:* AMNH 26820, partial skull with left and right M^2–M^3, jaw with left and right M_2, and skeletal fragments.

Horizon and localities: Late Pleistocene of North America. The type is from the Brunswick canal, Darien, Georgia. Other specimens from many localities.

Mammuthus primigenius (Blumenbach), 1803

Elephas primigenius Blumenbach, 1803: p. 697.
Elephas jacksoni Mather, 1838: p. 96, fig. A (p. 363).
Elephas americanus DeKay, 1842: p. 101, pl. 32, fig. 2.

Lectotype: Left lower M3 in the Blumenbach Collection of the Zoological Institute of the Museum of the University of Göttingen (see page 60).

Types of synonyms: *E. jacksoni:* mandible with left and right M_2–M_3, present whereabouts of type unknown to author. *E. americanus:* partial M_3; destroyed by fire.

Horizon and localities: Late Pleistocene of Eurasia and Wisconsin Glaciation of North America. The lectotype is from Siberia, locality unknown. North American specimens from many localities in Alaska, Canada and northeastern United States.

ELEPHANTINAE, *gen. indet.*

Several species of elephant have been described from southern Africa which cannot confidently be assigned to genus. Each is known only from a single, incomplete type specimen that cannot be placed into a stratigraphic or faunal context with accuracy. All are early forms, probably late Pliocene in age, and therefore fall into the period of time when several groups were passing through similar stages with respect to their molar teeth. They are all similar enough to be included within a single species and may be conspecific with *Loxodonta adaurora* or with *Mammuthus africanavus.* From the great lateral tapering of some of these molars, they seem closer to *M. africanavus,* but without a larger sample or, better, cranial evidence, a confident decision cannot be made. In a previous work I (1970b) considered the species listed below as *nomina dubia,* with the names applying only to the type specimens. This view is maintained here, at least until better material from southern Africa becomes available.

The four species names here in question are as follows:

Archidiskodon vanalpheni Dart, 1929. A single incomplete left M^3 from the Middle Terrace of the Vaal River, South Africa. There were probably about ten plates on the complete molar. The plates are broad and well separated, resembling those of *L. adaurora.*

Archidiskodon milletti Dart, 1929. An incomplete left M^3 from the Middle Terrace of the Vaal River. The plates are thinner than in *A. vanalpheni,* but not greatly. The sides of the molar taper towards the apex as in *M. africanavus.*

Archidiskodon loxodontoides Dart, 1929. A single fragmentary left M^3 from the Middle Terrace of the Vaal River. From what remains of this specimen it appears to be close to the previous two molars, but specific diagnosis is not possible.

Loxodonta griqua Houghton, 1922. Several very fragmentary specimens from the Vaal River gravels of Griqualand West, horizon unknown. This material is too incomplete to allow confident specific comparison with any other taxon. It was made the type species of a new genus, *Metarchidiskodon,* by Osborn (1934).

IDENTIFICATION OF ELEPHANT SPECIES

The identification of elephant species can often be an arduous task, especially with incomplete material. As seen in the preceding section, cranial criteria in many cases are more important for generic identification than are dental parameters. Because of the scarcity of skull material in the fossil record, such identifications must rely almost entirely on dental evidence. Such evidence is sufficient for the identification of most later species, but for many of the earlier forms other criteria must usually be employed. Geographic and geologic ranges can often provide valuable data in this regard. However, because these ranges are deduced partially from negative evidence, they cannot be considered as absolute.

In the following discussion I have attempted to point out the major features that may be used to identify elephant species from whatever fossil material may be available, except postcranial remains. Comparisons will be made primarily between those species most easily confused, but all species occurring in any given time period will be mentioned. At the end of the section an oversimplified key to the identification of fossil species will be given.

Latest Miocene (7.0–5.5 m.y.) The earliest elephantid species are African in distribution, first appearing in late Miocene deposits about 7 million years ago. During this period, the dominant elephants were the stegotetrabelodontines of which at least three more or less contemporary species are known. *Stegotetrabelodon syrticus* was North African in distribution and may be distinguished from other species by its very broad molar teeth with free columns in every transverse valley. The skull and jaws each contain a pair of very long and extremely narrow incisors. In the lower jaw these may extend 170 cm. beyond the very elongated symphysis while being less than 9 cm. in diameter. In the contemporaneous East African species, *S. orbus,* the molars are narrower throughout (see tables 2 and 3), free columns are present only in the anterior few valleys and the mandibular tusks are shorter with an exposed length of only about 60 cm. These are relatively more robust than in *S. syrticus.* In the third and as yet undescribed species from central Africa, mandibular tusks were lost in some individuals, but a very long symphysis remained.

Contemporaneous with *Stegotetrabelodon* in late Miocene deposits of Africa was the earliest elephantine taxon, *Primelephas,* known from two species. The major distinction between *Stegotetrabelodon* and *Primelephas* was in the mandible and anterior molar teeth. The symphysis of the latter was greatly reduced in length and only a small pair of incisors remained. These were widely separated in contrast to the narrowly spaced lower tusks of the stegotetrabelodontines. The third molar teeth of *Primelephas* lack the deep median cleft of *Stegotetrabelodon* and have thinner enamel (4 to 5.5 mm. *versus* 5 to 6.5 mm.). In M1 and M2 these features are even more striking. The M2 is narrow throughout, contrasting strongly with the short, broad M2 of *Stegotetrabelodon.* Isolated plates may be identified as *Primelephas* by the lack of a deep median cleft, and by the less triangular cross section than in *Stegotetrabelodon.* Of the two known species of *Primelephas, P. korotorensis* is distinguished from *P. gomphotheroides* by its relatively higher crowned molars. The hypsodonty index (HI) is 75 for the former and 55 to 65 for the latter.

Early Pliocene (5.5–3.5 m.y.): In the early Pliocene of Africa the stegotetrabelodontines were completely replaced by the Elephantinae; *Primelephas* was succeeded by three species, each at the base of a distinct phyletic complex. For this reason they have been placed in three genera, *Loxodonta, Elephas,* and *Mammuthus.* If cranial material is available *L. adaurora* may be recognized easily by its clearly *Loxodonta*-type of skull. The parietals are rounded, not expanded, tusk sockets are long and flaring, and the frontal region curves gently into the temporalis fossae. The molar teeth have only 9–11 plates with a hypsodonty index of about 100.

The contemporary species *Elephas ekorensis* may be distinguished on cranial characters by its laterally expanded parietals and occiput, by the small tusk sockets and the sharp fronto-parietal ridges separating the frontal region from the temporalis fossae. The molar teeth have one or two more plates than in *L. adaurora* (11–12) and are proportionately higher crowned (HI = 110–130). The plates are relatively thin and more parallel-faced. They are only rarely confused with the more thick-based molars of *Loxodonta.* The mandible is long in both species but the condyles of *Elephas* are generally elongated laterally, whereas those of *Loxodonta* are round.

The third species found in African deposits of this age is *Mammuthus subplanifrons.* The skull is unknown, but should be expected to resemble that of *Elephas* except for the fronto-parietal ridges. The upper tusks are twisted in typical *Mammuthus* fashion, and should be easily distinguishable from the nearly straight tusks of *Loxodonta* or *Elephas.* The molar teeth can be distinguished from those of *Elephas* by their lower crown height; the HI is less than 100 in all known specimens.

The number of plates on M3 is 7–9, and is lower than in both *E. ekorensis* and *L. adaurora,* although the latter may occasionally have as few as 8 plates on M3 (see tables 8 and 26). It is unlikely that *M. subplanifrons* would be confused with *Elephas;* the generally lower crown height and thicker enamel will help distinguish it from *Loxodonta.*

Late Pliocene (3.5–2.0 m.y.): During the later half of the Pliocene as defined here, *Loxodonta adaurora* persisted with little change in dental morphology. We might expect some change to have occurred in cranial structure toward those specializations found in the living species, but no *Loxodonta* skull material is known from this time period. *Elephas recki* had evolved from *E. ekorensis,* passing through two early stages during this period in East and Central Africa. Stage 1 is easily distinguished from *Loxodonta adaurora* by its greater number of plates (12–14), its greater HI (125–135), its thinner enamel (2.8–3.4 mm.), and especially by its very thin, closely spaced plates with a LF of greater than 4.5 (see tables 8 and 13). Stage 2 *E. recki* is even more distinct from *Loxodonta.* It differs from stage 1 in having a proportionately greater HI (125–155), a LF generally in excess of 5.0, and in having broadly folded enamel even in early stages of wear. In stage 1, the enamel is not folded except near the base of the crown. Both stages of *E. recki* have prominent posterior enamel loops, as does *L. adaurora,* but these may be quite small or even absent in *E. recki* stage 2.

Two additional species were present during this time period, but both are predominantly North African in distribution. *Mammuthus africanavus* was an early species of mammoth. The skull bears long, massive, and spirally twisted tusks and the occiput is laterally expanded as in later species of the genus. The molars may cause some confusion, however, in that they are rather similar to those of *L. adaurora.* However, these two species seem to have been geographically well separated, *M. africanavus* in Chad and North Africa, and *L. adaurora* in eastern and southern Africa. As new localities are discovered this distinction may disappear. There are no clear-cut metrical differences between the molars of these two species. The LF is slightly greater in *M. africanavus* and the HI of M_3 generally is higher (see tables 8 and 28). The sides of the plates taper strongly towards the apex in the mammoth, but much less so in *Loxodonta.*

The last stage of the *Mammuthus* lineage in Africa was *M. meridionalis* which appears to have been exclusively North African in distribution. It may be easily distinguished from the earlier *M. africanavus* by its higher HI which is generally 140–160, and by its greater number of plates, 12–14 on M3. Although *E. recki* never reached North Africa, its molars may be confused with those of *M. meridionalis.* In the absence of cranial evidence, which would always clearly distinguish these two forms, the only dental feature that is useful in separating them is the greater breadth of the mammoth molar and the general lack of a median enamel loop. The latter character has been replaced by a broadening of the plate in its median portion.

In Europe the only elephant of this period was *M. meridionalis,* first appearing as the Laiatico Stage, and evolving into the more progressive Montovarchi Stage before the close of the Pliocene. The only Asiatic elephant of this time was *Elephas planifrons*. The molars of *planifrons* are difficult to separate from those of the Laiatico Stage of *M. meridionalis,* but on cranial evidence, the two cannot easily be confused. The skull of *E. planifrons* is typically *Elephas* with sharp frontoparietal ridges and nearly straight tusks. It also differs in having a flat frontal region in contrast to the concave frontal area of *M. meridionalis.*

Early Pleistocene (2.0–1.0 m.y.): *Loxodonta adaurora* does not survive into this period. A descendant species transitional to *L. africana* must have existed, but no such form is known at present. In Africa, another species of *Loxodonta, L. atlantica,* appears and is easily distinguished by its highly folded enamel, lozenge-shaped enamel figures and often bifurcated median enamel loops. The most common species found in Africa at this time is *Elephas recki*. There is little room for confusion here as this form is distinct on many characters. The skull of *E. recki* is anteroposteriorly compressed and has a laterally expanded parietal region, concave dorsal crest and frontal region, and long, parallel and closely spaced tusk sockets. The molar teeth have simple median loops, folded enamel and are never lozenge-shaped as in *L. atlantica.* The crown height in the latter is generally greater than in *E. recki* (see tables 11 and 14).

Mammuthus meridionalis is the only elephant in Europe during this time. In Asia, *E. planifrons* persists into the Pleistocene with some change in molar characters, generally attaining a greater number of plates and becoming higher crowned than its earlier stage. The roughly contemporaneous species, *E. hysudricus,* appears to have coexisted with *E. planifrons.* The two forms are generally distinguishable although some specimens may be difficult to place. *E. hysudricus* molars have 12–14 plates, in contrast to the 8–13 of *E. planifrons*. The lamellar frequency is also greater as is the HI which is probably the best means of separating these two species (see tables 20 and 23). When cranial material is available, these two forms are unmistakable. *E. hysudricus* is easily recognized by its greatly expanded parietals, concave frontal area, and extremely narrow intertemporal region. The only other Asiatic species in the Early Pleistocene is *E. celebensis.* This is a dwarf species known only from Celebes and Java and is half the size of *E. planifrons.* It is easily identified because of its small size, but otherwise is *planifrons*-like in dental features. Care must be taken, however, to identify properly the molars as to serial position so as to avoid making comparisons between anterior molars of *planifrons* and posterior molars of *celebensis*.

During the latest phases of the early Pleistocene, *M. meridionalis* entered North America where it was the only elephant until well into the middle Pleistocene.

Middle Pleistocene (1.0–0.2 m.y.): In Africa, three contemporaneous species were present during this time period. *Loxodonta atlantica* and *L. africana* are easily distinguished by the highly folded enamel and bifurcated median enamel loop in the former, and smooth enamel with simple loops in the latter. Stage 4 of *E. recki* is the dominant elephant in East Africa during the earlier part of this time period. It is distinguished from either species of *Loxodonta* by its lack of lozenge-shaped enamel figures and the absence of enamel loops. If cranial material is available, identification of this species can be certain.

In Europe, *Mammuthus armeniacus* succeeded *M. meridionalis*. It is distinguished from the contemporaneous *Elephas namadicus* by its parallel-sided enamel figures with only a slight, rounded median expansion and thin, wavy or ribbed, but not folded enamel. In contrast, *E. namadicus* has intensely folded enamel with small pointed median enamel loops. In the skull, *M. armeniacus* has the twisted tusks typical of the mammoths. The parietals are rounded and high with a long, slightly concave frontal region. Tusk sockets are long, parallel, and closely spaced. The skull of *E. namadicus* is low and rounded with sharp fronto-parietal ridges, a short frontal region and a strong, overhanging frontal crest. The pygmy elephant of Sicily, *E. falconeri*, is not confused with any of the mainland species, but is not easily separated from specimens referred to other Mediterranean dwarf species. Confusion will persist until this group is critically reexamined.

Elephas platycephalus is presumably of middle Pleistocene age in southern Asia and should be clearly distinguishable on cranial characters which differ widely from those of *E. namadicus* and *E. hysudrindicus*. Here, the skull is low and elongated anterioposteriorly, with a rounded, convex frontal region and non-expanded parietals.

The North American middle Pleistocene elephant is *Mammuthus imperator*. This species is easily distinguished from the earlier *M. meridionalis* by its greater number of molar plates (16–19) and its higher hypsodonty index (125–220). The mandibular rami are more robust than in any other North American species.

Late Pleistocene (0.2 m.y.–Recent): Two elephants coexisted in Africa during this time, *Loxodonta africana* and *Elephas iolensis*. The very large molars of *E. iolensis* are not easily confused with those of *L. africana* and are readily recognized by their great hypsodonty index (150–300), their intense enamel folding and their lack of median loops. The lozenge-shaped, smooth enameled molars of *Loxodonta* should offer no problem in identification.

Only two species are present in Europe during this period, *Mammuthus primigenius* and *Elephas namadicus*. As was the case in the middle Pleistocene, the mammoth is not easily confused with *Elephas*. The cranial and dental specializations seen in *M. armeniacus* are carried to a greater extreme in *M. primigenius*. In southern Asia, only *Elephas maximus* is known, although *E. namadicus* may have survived in some areas, especially in Japan. *M. primigenius* was the only species in northern Asia at this time and in North America it was contemporaneous with *M. columbi*. The latest stage of *M. columbi* is often confused with *M. primigenius* on molar characters alone, but may be distinguished by its broader, longer skull in which the occiput is more rounded and the tusk sockets are distally diverging.

The following key to the species of Elephantidae attempts to provide an oversimplified method for the identification of fossil species. Since molar teeth are the most commonly found remains of these animals, the key relies heavily on molar parameters. The scheme also utilizes cranial characteristics which are often the only certain means of specific identification. It must be kept in mind that in many cases, individual specimens will be morphologically intermediate between two defined taxa and positive reference to either one may be impossible. The key is designed to work best with samples of specimens in which average measurements and some idea of variability can be obtained.

KEY TO THE SPECIES OF ELEPHANTIDAE

1.*a*. Geographic distribution African 2
  b. Geographic distribution European 13
  c. Geographic distribution Asiatic 16
  d. Geographic distribution North American 21

2.*a*. Molar teeth with deep median cleft, jaw with long symphysis .... (Stegotetrabelodontinae) 3
  b. Molar teeth lacking deep median cleft, jaw with short symphysis .... (Elephantinae) 4

3.*a*. Third molars greater than 110 mm. in width, posterior columns behind every plate, mandibular tusks 55–60% of jaw length ....*S. syrticus*
  b. Third molars less than 110 mm. in width, posterior columns behind first few plates only, mandibular tusks 35–40% of jaw length ....*S. orbus*

4.*a*. Third molar teeth with 7–8 plates or less, jaw with small mandibular incisors ....(Primelephas) 5
  b. Third molar teeth with 8 or more plates, jaw lacking incisors 6

5.*a*. Relative crown height 65% or less of width ....
.................................*P. gomphotheroides*
  b. Relative crown height 75% of width .....*P. korotorensis*

6.*a*. Third molar with smooth enamel 7
  b. Third molars with folded enamel 11

7.*a*. Upper tusks spirally twisted 8
  b. Upper tusks not spirally twisted 10

8.*a*. Third molars with 8–11 plates, crown height less than 115% of width 9
  b. Third molars with 11–14 plates, crown height greater than 120% of width ....*M. meridionalis*

9.*a.* Third molars with 8–9 plates, relative crown height 65–90% of width ....*M. subplanifrons*
  b. Third molars with 9–11 plates, relative crown height 90–115% of width ....*M. africanavus*
10.*a.* Third molars with 9–11 plates, relative crown height less than 110% of width ....*L. adaurora*
  b. Third molars with 11–12 plates, relative crown height 110–125% of width ....*E. ekorensis*
11.*a.* Median enamel loop bifurcated, plate expanded in midline ....*L. atlantica*
  b. Median enamel loop simple, plate not expanded in midline 12
12.*a.* Crown height of third molars, 80–160 mm. ....*E. recki*
  b. Crown height of third molars, 160–250 mm. ....*E. iolensis*
13.*a.* Enamel highly folded with median expansion, skull with frontal crest, tusks straight ....*E. namadicus*
  b. Enamel smooth to folded, lacking median expansions, skull lacking frontal crest, tusks twisted 14
14.*a.* Third molars with 11–14 plates, enamel unfolded, relative crown height 100–150% of width ..*M. meridionalis*
  b. Third molars with more than 14 plates, enamel folded..15
15.*a.* Third molars with 14–21 plates, enamel 1.5–3.0 mm. thick, LF = 5–8 ....*M. armeniacus*
  b. Third molars with 20–27 plates, enamel 1.3–2.0 mm. thick, LF = 7–11 ....*M. primigenius*
16.*a.* Enamel unfolded, third molars with 8–13 plates, relative crown height less than 115% of width 17
  b. Enamel folded, third molars with 12–27 plates, relative crown height greater than 115% of width 18
17.*a.* Third molars more than 180 mm. in length, LF less than 6, jaw lacking tusks ....*E. planifrons*
  b. Third molars less than 180 mm. in length, LF greater than 6, jaw usually with tusks ....*E. celebensis*
18.*a.* Enamel highly folded with pointed median expansions, skull with frontal crest ....*E. namadicus*
  b. Enamel folded, lacking median expansions, skull lacking frontal crest 19
19.*a.* Third molars with 12–17 plates, relative crown height less than 170% of width, enamel moderately folded with median loop ....*E. hysudricus*
  b. Third molars with 18 or more plates, relative crown height greater than 170% of width, enamel highly folded, lacking median loops 20
20.*a.* Third molars with 18–20 plates, relative crown height less than 200% of width, skull with laterally expanded and dorsally flattened parietals ....*E. hysudrindicus*
  b. Third molars with 23 or more plates, relative crown height greater than 200% of width, skull lacking dorsal flattening of parietals ....*E. maximus*
21.*a.* Third molars with 20 or more plates, lamellar frequency 7.0–12.0 22
  b. Third molars with less than 20 plates, lamellar frequency 4.0–6.0 23
22.*a.* Skull with very high, pointed and expanded parietal region, rounded occiput, distally parallel tusk sheaths ....*M. primigenius*
  b. Skull with rounded parietal region, flat occiput, distally flaring tusk sheaths ....*M. columbi*
23.*a.* Third molars with 16–18 plates, maximum crown height 150 mm. or more ....*M. imperator*
  b. Third molars with 10–14 plates, maximum crown height 150 mm. or less ....*M. meridionalis*

V. CORRELATION OF PLIOCENE/PLEISTOCENE DEPOSITS

Faunal correlation on any taxonomic level must be approached with the greatest of caution, especially when dealing with areas between traditionally separate, though ephemerally continuous, biogeographic regions. Ideally any such correlation should be based on total faunal similarity and should take into account available geological evidence. In actual practice this is often difficult or impossible because of incomplete data, and reliance on several well-documented groups usually is necessary. Within the Pleistocene epoch, groups that were undergoing rapid evolution and geographic expansion are particularly useful in this respect; others seem to have remained nearly unchanged through long periods of time and to have been more limited in their geographic distribution. Some groups, such as the Proboscidea, underwent bursts of evolutionary activity and geographic expansion during transitions between adaptive zones and subsequent post-adaptational shift modifications. These evolutionary phases are often accompanied by rapid branching of phyletic lines in an adaptive radiation. Such groups provide unusual opportunities for correlation on both local and intercontinental scales.

The use of "index" or "guide" fossils, especially when applied to mammals, is of dubious value for precise correlation. The first appearance of the genera *Elephas* (*s.l.*), *Equus* (*s.l.*) and *Leptobos* has long been used as indicating the opening phases of the Villafranchian in Europe and Asia. The tacit assumption that any group of mammals can and does appear suddenly and spread throughout the world rapidly enough to be considered synchronous has not adequately been demonstrated. *Elephas* (*s.s.*), for example, arose in Africa during the early Pliocene and extended its range to Asia in the late part of that epoch. It does not appear to have reached Europe until the Cromerian, middle Pleistocene. The typical Villafranchian elephant of Europe was an early *Mammuthus* derived from North Africa, presumably via a different route from that followed by *Elephas* to Asia (see section IX). Although roughly contemporaneous geologic events, whatever these may have been, probably allowed both these expansions to occur across different land corridors at approximately synchronous times, the sudden appearance of elephantine proboscideans in northern continents is clearly a more complex phenomenon than previously believed. The danger of uncritically using "first appearances" as indices for intercontinental correlation, especially before the systematics has been clearly worked out, becomes even more evident when we realize that *Elephas* as a definable genus existed in Africa for some two million years before its first occurrence in Eurasia.

Evidence suggests the existence of *Equus* in North America as early as the early Blancan (Skinner, pers. comm.). In Africa this genus is now definitely recorded in late Pliocene deposits, and its presence in even earlier deposits is indicated. Thus, the use of index fossils must take into account centers of origin, rates of evolution and directions and rapidity of expansion of the groups utilized. Until these factors are known, it is not likely that mammalian "guide" fossils will prove very useful for correlation purposes.

An additional major problem lies in the identification of specimens. Few paleobiologists can examine in detail material from widely scattered localities when working on local faunas. As a result of this, the number of supposedly endemic taxa of Pleistocene mammals recorded within each major geographic area is staggering. This is in part due to the lack of adequate interregional comparisons. Without secure identifications based on comparisons of fossils from widely separated areas, it becomes hopeless to attempt precise intercontinental correlations on any level below the generic, and usually even this is questionable. There is a great need for more monographic studies of taxa over their entire ranges.

Although the problems are numerous, even a rough correlation is necessary if evolutionary inferences are to be drawn from paleobiological material. Without such inferences, the study of fossils becomes rather academic. Therefore, an intercontinental correlation of late Miocene to Pleistocene deposits is presented here (fig. 12) as a synthesis drawn partly from my own work and partly from the work of others. Absolute radiometric control points within this time span are, fortunately, becoming ever more numerous. We are a long way yet from having anything like a complete time scale but, in the meantime, the available dates serve as extremely useful datum points and as checks upon correlations based on faunal evidence. (Reciprocally, of course, securely correlated faunal assemblages serve to check the reliability of radiometric dates.) Faunal similarity, usually on the generic level, has been heavily relied on here, but I have not been able to take into account ecological distinctions, which in most cases are uncertain. In spite of all the problems, it is possible to draw a broad correlation which allows analysis of morphological data in a temporal framework.

Cooke (1967) has emphasized the need for an adequate and acceptable terminology for Pleistocene subdivisions and has pointed out some of the existing problems. His attempt to set up a relative sequence of "faunal spans" in southern Africa is useful for that restricted area, and is a hopeful step toward the establishment of a biostratigraphic sequence in Africa. It cannot provide the much-needed nomenclature that would be useful on a continental, let alone an intercontinental scale. The South African Pleistocene fauna contains a significant number of endemic genera that make a faunal terminology based on them unsuitable for other areas.

I do not propose to designate formal biostratigraphic zones, as this should be undertaken only by a committee of geologists and paleobiologists involved with these problems. Nor do I intend to place the controversial Pliocene/Pleistocene boundary within the stratigraphic sequence. The solution to the latter problem will not be at hand until authors cease trying to "recognize" the "elusive" boundary as if it were a universal and absolute horizon clearly labeled in the field. The Tertiary-Quaternary boundary must be subjectively set on absolute radiometric and possibly paleomagnetic criteria that can then be calibrated with faunal zones. This is not to say that the boundary will not be a real entity, for, once defined, it will be as real as any objective division.

For purposes of convenience, however, the terms Pliocene and Pleistocene as used here must be defined, as it is impossible to avoid their use. A rough Pliocene/Pleistocene boundary is placed at approximately 2.0 million years. Time periods earlier and later than 2.0 m.y. are designated as Pliocene and Pleistocene respectively. The boundary point is referred to here simply as Pliocene/Pleistocene. Under this scheme the European Villafranchian is divided, with Mount Coupet emerging as earliest Pleistocene, and earlier Villafranchian deposits such as Roca Neyra as latest Pliocene.

AFRICA

In figure 12 the African sequence is taken back to the latest Miocene because of the early appearance on that continent of the Elephantidae. A less comprehensive correlation of African deposits of this time period has already been published (Maglio, 1970a) and the present arrangement, although basically similar, presents several points of refinement.

The North African sequence has been summarized by Arambourg (1949, 1952), Howell (1959, 1960) and Cooke (1964), and a monograph on the North African Pleistocene by the late Professor Arambourg has just appeared (1969–1970). Although these localities share many genera with those in East Africa, there are only a few species that occur both north and south of the Sahara.

The age of the Sahabi deposits in southeastern Libya is now considered to be latest Miocene (middle Pliocene of Aguirre, 1969 and Maglio, 1970a) and not as old as originally believed (Petrocchi, 1941). The presence of a primitive *Nyanzachoerus* (Cooke, pers. comm.), *Hippopotamus, Brachypotherium* (Hooijer and Patterson, 1972) and a primitive stegotetrabelodont elephant confirms this conclusion. The exact position of the Sahabi fauna within the Mio/Pliocene sequence cannot be positively established, but from analogy with East Africa it appears to be close to and perhaps slightly older than that from Lothagam 1.

Ichkeul, Aïn Boucherit, Fouarat, and a new site at Kebili in Tunisia (Arambourg, 1969–1970) fall into Arambourg's early Villafranchian [14] stage and are all

---

[14] Application of the term Villafranchian within Africa has led to confusion in the past. It is used here for deposits equivalent in age to those of the European Villafranchian as recently defined by Azzaroli (1970). It thus includes Etouairés at its base and the Weybourne Crag and Sainzelles at the top (see fig. 12).

roughly comparable to each other in age. Direct comparison with East African faunas is difficult, but the presence of *Stylohipparion, Equus, Ceratotherium, Mesochoerus,* and Recent species of *Hyaena* and *Gazella* indicate a fauna of the middle to later Shungura (Omo Valley) type. The advanced short-jawed gomphothere, *Anancus,* appears to have survived in North Africa for some time after its disappearance from other parts of that continent. If we assume that *Equus* entered North Africa before spreading southward, then its presence here somewhat earlier than in East Africa would not be surprising. It must also be remembered that the genus *Equus,* as it applies to African fossils, is still not adequately understood. The general faunal aspect of these North African deposits is comparable to that of the Shungura and later Kaiso Formations in East Africa (see below).

The later faunas from Ternifine and Sidi Abder Rahmane lack the characteristic genera of the early Pleistocene in Africa. Living species of hyaena, giraffe, lion, and gazelle are present. Two elephants, a specialized *Loxodonta* and a late member of the *Elephas* lineage in Africa, are present but do not occur together in the same deposits, suggesting some degree of ecological isolation. These faunas are middle Pleistocene in age, with the Sidi Abder Rahmane and Rabat-Casablanca deposits probably being post-Olduvai Bed IV.

The East African sequence presently provides the standard against which correlation within the continent must be made. Not only are the faunas generally well known, but the radiometric calibration available for Olduvai Gorge, Omo, Kanapoi, and the Lake Baringo area provide an absolute scale of reference.

Scattered and poorly known faunas from around Lake Baringo in Kenya give us our only glimpses of the late Miocene of Africa. Two important but very fragmentary faunas have been recovered, one from Mpesida (approximately 7 m.y.) and the other from Kaperyon dated at about 5 m.y. (W. W. Bishop, pers. comm.). Although the Kaperyon and Mpesida faunas are not extensive, they are comparable to the Lothagam 1 fauna and serve to support the dating of the latter.

The best latest Miocene fauna known at present from East Africa is that of Lothagam 1 (Patterson *et al.,* 1970). The fauna contains the most primitive known elephants (*Stegotetrabelodon* and *Primelephas*), the last survivor of the brachypotherine rhinoceroses (Hooijer and Patterson, 1972), a very primitive hyaenid, early suids (Cooke and Ewer, 1972), and very primitive bovids (A. Gentry, pers. comm.). This fauna is clearly older than those of the Kanapoi and Mursi Formations which are dated at about 4.0 m.y. An intrusive basalt, post-dating deposition of the uppermost unit, Lothagam 3, is dated at 3.71 ± 0.23 m.y., thus giving an upper limit for deposition of the entire Lothagam sequence. A lower limit of 8.31 ± 0.25 m.y. has been obtained for an underlying basalt flow (Patterson *et al.,* 1970). From faunal differences between Lothagam 3 and 1, it seems unlikely that Lothagam 1 could be less than 5.0 m.y. in age, and it is probably closer to 5.5 or 6.0 m.y.

On the east side of Lake Rudolf, recent work has yielded an early fauna in the area around Kubi Algi near Allia Bay. The fauna is as yet only partially known, but contains *Hippopotamus immagunculus, Gomphotherium* sp., *Nyanzachoerus* sp., *Notochoerus capensis,* and the earliest known stages of *Loxodonta adaurora* and *Elephas ekorensis* (Maglio, 1971). A K/Ar age of 4.5 ± 0.1 m.y. has recently been obtained for a pumiceous tuff immediately and conformably underlying the fossil-bearing deposit (J. A. Miller and F. Fitch, pers. comm.).

The next absolute reference points in the time scale are the Kanapoi and Mursi Formations. A basalt overlying the Mursi has yielded an average age of 4.05 ± 0.20 m.y. (Brown and Lajoie, 1970). The Kanapoi sediments are overlain by a basalt that is now known to be about 4.0 m.y. old (F. Fitch, pers. comm.). (Also see Maglio, 1970, and Patterson *et al.,* 1970, for discussions of the age determination of Kanapoi.) These faunas show a distinctly progressive aspect as compared with that of Lothagam 1 with many new forms appearing, and many old ones no longer present. A more advanced elephant of the genus *Loxodonta* has replaced the more archaic elephants of Lothagam 1, and a very early form of the White Rhinoceros, *Ceratotherium,* is present. More progressive species of *Nyanzachoerus* and an early *Notochoerus* (*N. capensis*) make their appearance here, and bovids typical of later deposits are common elements. *Anancus* is still present as a rather progressive species different from, and probably unrelated to, the more generalized North African *Anancus osiris. Equus* is unknown from deposits of this age, but *Stylohipparion* is common. The Lothagam 3 fauna is identical to that from Kanapoi, and these beds at Lothagam may be no more than a different facies of the depositional events that produced the Kanapoi beds forty miles further south.

In a review of the "Laetolil" (*sensu lato*) Proboscidea, I (1969) recently suggested that the fauna from several southern Serengeti sites then known under that name was mixed, containing a Pliocene, Kanapoi-like assemblage, and a later group of upper Shungura aspect. Since that time it has been suggested that the name "Laetolil" should be applied only to the single site originally worked by Leakey (L. S. B. Leakey, pers. comm.). I adopt this suggestion. Although the name Laetolil has been used in the less restricted sense by almost every author since Kent (1941), it seems possible that the various localities near the Vogel River, all referred to as Laetolil by Dietrich (1942), are not entirely the same as the original locality. The latter has yielded material referable to the more recent fauna, whereas the former localities consist primarily of the

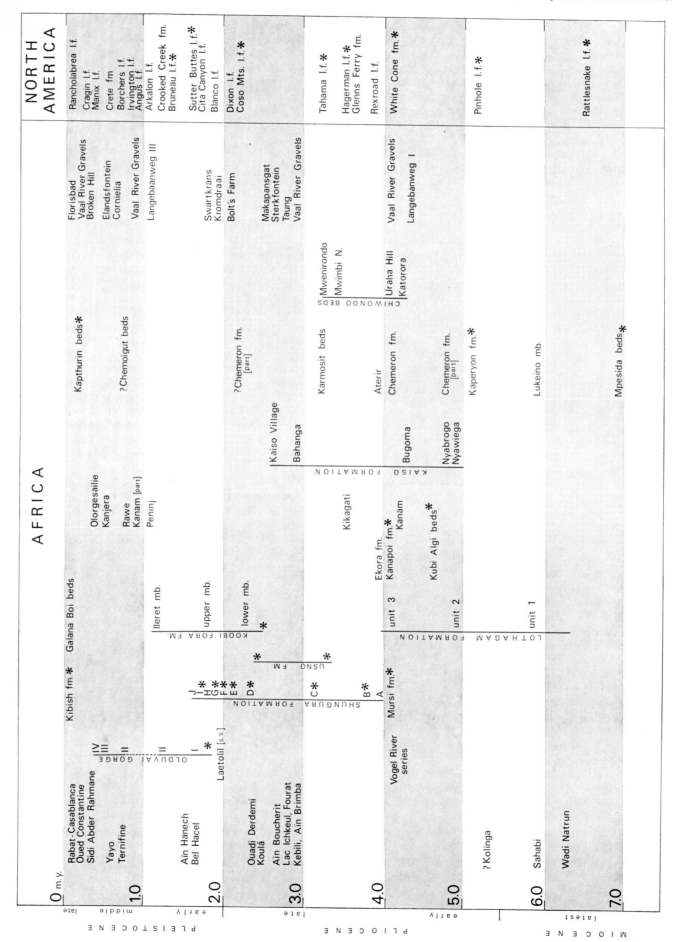

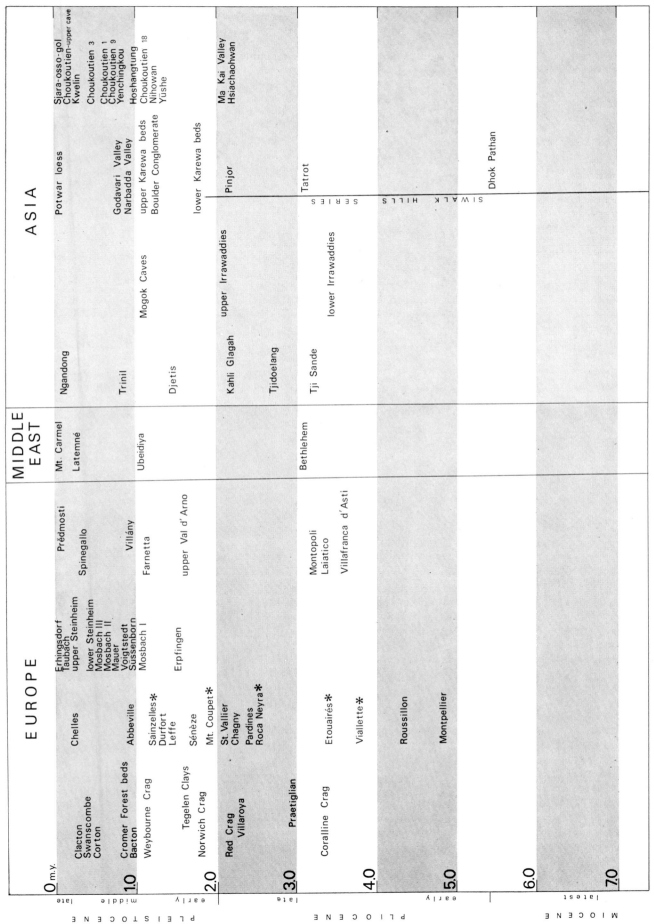

FIG. 12. Proposed correlation of major late Miocene to Pleistocene deposits of Africa, Europe, Asia, and North America based on faunal and radiometric data. Solid lines indicate known continuous sequences. Broken lines show presumed extensions of stratigraphic sections. Asterisks indicate the positions of currently available potassium/argon age determinations used as control points for absolute calibration.

older fauna which may in part be reworked and mixed with later elements. To avoid further confusion, the name Vogel River Series, as used by Bishop (1967), is applied to these Serengeti localities other than Leakey's original site.

As discussed elsewhere (Maglio, 1970a), one of the elephants from the Vogel River beds is identical with *Loxodonta adaurora* from Kanapoi. The other species from the presumed later fauna and from Laetolil proper should be referred to *Elephas recki* from the upper part of the Shungura Formation in the Omo Valley. Since that time, *L. adaurora* has been found as late as Bed G in the Omo beds and the lower member at Koobi Fora. In view of these finds it seems possible that the early age proposed for part of the Vogel River Series may have been excessive. Ecological conditions during deposition of the Vogel River beds may have favored the exclusive presence of *L. adaurora* and the exclusion of *E. recki*. If this was so, then the Vogel River beds could be younger than those at Kanapoi. Nevertheless, the clear differences in preservation of these two taxa make it likely that we are dealing with two different depositional situations. The problem of the age of the fauna containing *L. adaurora* certainly requires further study.

In a recent review of the Kaiso Formation, Cooke and Coryndon (1970) demonstrate the early nature of the faunas from Nyawiega and Nyabrogo. These localities include the lowest part of the Kaiso Formation. The elephant is close to the Lothagam 1 *Primelephas* and the fauna as a whole indicates that at least part of the Kaiso Formation correlates with the Kanapoi-Mursi to Lothagam 1 time span.

The upper section of the Kaiso Formation at Kaiso Village contains typical late Pliocene elements as does the Shungura Formation of the Omo basin. *Ceratotherium simum* and an early stage of *Elephas recki* are present. The identification of a fragmentary specimen as *Equus* (Cooke and Coryndon, 1970) is questionable; it is more likely *Stylohipparion*.

The fauna from the base of the Chemeron Formation near Lake Baringo in Kenya is clearly of early Pliocene age, but examination of the faunas from the various localities indicates a longer span of time for this formation than has been previously believed. On faunal evidence, most of the Chemeron deposits would seem to range from a late Lothagam 1 age equivalent to that of Ekora (i.e., from approximately 5.0 m.y. to 4.0 m.y.). One site, however, requires further discussion. Locality J.M. 90 has yielded a fauna that can not be contemporaneous with the main Chemeron fauna. The elephant is a stage 2 *Elephas recki* like that from the upper Shungura Formation. A subspecies of the living *Ceratotherium simum* is known from very complete material (Hooijer, 1969), and there is a colobine monkey described by R. Leakey (1969) as *Paracolobus baringensis* of uncertain affinity. None of the earlier forms common to other "basal" Chemeron localities are known from J.M. 90 (A. Hill, pers. comm.). This situation certainly requires geological confirmation, but on present faunal evidence this locality seems to correlate best with the upper part of the Shungura Formation (Cooke and Maglio, 1972).

In the initial publications on the fossiliferous deposits at Koobi Fora on the east side of Lake Rudolf the age of the lower member of the Koobi Fora formation was inferred to be close to 2.6 ± 0.26 m.y., which is the date obtained for a pumice in a water-laid tuff low in the section (Leakey, 1970; Behrensmeyer, 1970; Fitch and Miller, 1970). More recent faunal evidence supports this as a lower age limit with a probable age range for this lower unit of between 2.0 and 2.6 m.y. (Maglio, 1972a). This correlates roughly with the Shungura Formation between tuffs D and I (Maglio, 1971).

Equus is present throughout the Koobi Fora sequence. This genus does not appear elsewhere in sub-Saharan Africa below the level of tuff G in the Shungura Formation. The Koobi Fora *Mesochoerus*, *M. limnetes*, is a more progressive stage than that which is typical of the earlier Shungura Formation. Specimens from the upper part of the Koobi Fora succession are closer to what has been called *M. olduvaiensis* from lower Bed II at the Olduvai Gorge. The elephant, *E. recki*, from the top of Koobi Fora is variable, but generally closer to stage 3 from Olduvai Beds I and lower II. The Shungura species *Hippopotamus protamphibius* has not been identified at Koobi Fora. Rather, a larger and more highly specialized species is present throughout the sequence (Coryndon, pers. comm.). A suid close to, but more primitive than, "*Afrochoerus*" *nicoli* appears at the top of the Koobi Fora sequence, suggesting an age comparable to that of Olduvai Bed I or lower Bed II.

About thirty miles north of Koobi Fora are the fossiliferous deposits of Ileret. Faunal correlation indicates near contemporaneity of the lowest beds at Ileret and the upper part of the Koobi Fora succession (upper member). Both are about Olduvai Bed I equivalent in age. The upper beds at Ileret (Ileret member) are younger than this on faunal criteria, but do not appear to be as young as Olduvai Bed IV. The pigs and elephants are close to, but more primitive than those from Bed IV (Maglio, 1972a).

Initial consideration of the Langebaanweg locality in Cape Province correlated the fauna of Horizon I with the Sterkfontein Faunal Span of Cooke (Cooke, 1967; Hendey, 1969). That correlation was based on the identification of a number of bovid genera, especially *Kobus*, *Hippotragus*, *Parmularius*, and *Madoqua*, but the identifications must be considered tentative (Q. B. Hendey, pers. comm.). The supposed presence of *Equus* was also a major consideration in arriving at this correlation. However, the specimens so referred are worn milk molars of *Hipparion;* the latter often resemble *Equus* when they are in late stages of wear. The only true *Equus* is a surface find from Baard's Quarry and, judging from its preservation, was derived from more recent, superficial deposits (Q. B. Hendey, pers. comm.). The elephant from the base of the section is now known to be an early grade of *Mammuthus subplanifrons* (Maglio and Hendey, 1970) and not *Stegolophodon* as originally described.

Placement of the Langebaanweg deposits in the middle Pleistocene (Cromerian equivalent) was also based on the relationship of these sediments to the 50-meter-high sea level in the Cape region. This was tentatively correlated with the 50–60-meter (Milazzian) level of Europe. However, it is doubtful that sea-level correlations of this type can be applied usefully over such great intercontinental distances (see also Van Bemmelen, 1950) especially in view of the great discrepancies even within such small areas as the Mediterranean basin.

On reevaluation, the fauna of Horizon I at Langebaanweg seems best correlated with those of Kanapoi and Mursi of the East African Pliocene sequence (Hendey, 1970). However, recent evidence indicates that a later fauna (in Horizons III and IV) is also present and is of an Olduvai Beds I-lower II age.

The geology of the Vaal River gravels has been reviewed by Cooke (1947, 1949). Although the stratigraphy is largely unknown owing to the concentration of fossils and gravel during a period of eluviation, the oldest mammalian assemblages contain typical early Pliocene elements of the Kanapoi type. Higher levels in the "Younger Gravels" contain more progressive faunas including *Phacochoerus* and *Elephas iolensis*, known elsewhere in middle to late Pleistocene deposits. The faunas of the Vaal River Gravels, then, would appear to cover a significant period of time.

The relative dating of the South African dolomite cave deposits in the Transvaal has been discussed by numerous authors, with few positive conclusions being reached (e.g., Cooke, 1964; Ewer, 1957, 1967; Howell, 1957). Geological analysis by Brain (1959) seems to have established a good relative sequence, with Sterkfontein and Makapansgat being older than Swartkrans and Kromdraai. This view has won general acceptance although the story within each of the two groups may be more complicated, and the faunal evidence is, in general, rather weak.

These deposits share numerous taxa, but many of these are endemic to southern Africa and hence not useful for correlation outside of that area. When only non-endemic genera are examined it is clear that none of these sites differs strikingly from the others in faunal composition, although in terms of overall similarity, Swartkrans seems closer to Kromdraai and to Bolt's Farm than to Sterkfontein and Makapansgat (see also Ewer, 1957). The differences are slight and in part may be ecological. Comparisons with localities outside of southern Africa show more significant results. This must be approached with caution, however, because of the paucity of small mammals from localities such as Omo, Koobi Fora, and Kanapoi, and the relative abundance of such microfauna in the Transvaal cave deposits. Nevertheless, the total faunas of Kromdraai and Swartkrans seem to correlate better with the upper Shungura (G) to Olduvai Beds I-II time range than with those of Olduvai Bed IV—Elandsfontein or Kanapoi—Langebaanweg. The general aspect of the Sterkfontein and Makapansgat faunas, especially the suids and the elephant, is like that of the earlier Shungura Formation (C-D), but the presence of several primitive elements in the Makapansgat fauna, associated with differences in preservation, suggest a mixed assemblage, with the older being perhaps 3.0 m.y. or more in age (H. B. S. Cooke, pers. comm.).

Although this correlation is provisional, it seems unlikely on faunal evidence that any of the known South African cave deposits could be older than about tuff C of the Shungura Formation. Nor does it seem likely on present evidence that any of them could be younger than lower Bed II at Olduvai Gorge, and certainly not as young as the Mindel glacial phase in Europe as suggested by Kurtén (1957, 1959).

EUROPE AND THE MIDDLE EAST

Faunal correlation of European Pleistocene deposits has been discussed recently by Flint (1957), Kurtén (1959, 1963, 1968), Zeuner (1959), Hürzeler (1966), Azzaroli (1970), and others. Although details of opinion differ, the consensus suggests an arrangement similar to that shown in figure 12.

Azzaroli (1970) has provided a faunal basis for subdivision of the Villafranchian in Europe and has reviewed the major European localities. Kurtén (1963) has discussed the faunas of many early and middle Pleistocene deposits. It is, therefore, not necessary to repeat here the evidence bearing on the present correlation, which draws heavily on the works cited above. Only those localities on which there is not yet complete agreement will be discussed here.

Kurtén (1963, 1968) suggested a pre-upper Val d'Arno age for Sénèze because of the presence of a primitive *Ursus etruscus* and of *Leptobos stenometopon*. Azzaroli (1970) has questioned these determinations and emphasized the presence of four genera, seemingly confined to his *subzone b* of the Upper Villafranchian. The elephant from Sénèze, *M. meridionalis* (Montavarchi Stage), does not differ from that of the upper Val d'Arno (see page 56 above), thus supporting Azzaroli's conclusion.

The early Villafranchian age of Villaroya in Spain (Heintz, 1967) has been questioned on faunal grounds. Indeed, the presence of such typically middle to late Villafranchian species as *Dicerorhinus etruscus*, *Leptobos etruscus*, and *Gazellospira torticornis* would seem to indicate a later age, and I follow Azzaroli (*op. cit.*) in considering this fauna as of later Villafranchian age.

The supposed post-upper Val d'Arno, or latest Villafranchian, age of the Tegelen Clays (Zeuner, 1959; Kurtén, 1963) has also been questioned by Azzaroli, who considers the Tegelen fauna to be typical of the Upper Villafranchian. In general the faunal evidence would seem to support this view, although a later age cannot yet be ruled out.

The age of the Erpfingen cave deposit was originally referred to the later Villafranchian by Lehmen (1953, 1957), but Azzaroli suggests a composite age from middle to post-Villafranchian because of the presence of *Cervus ramosus* and *Megaloceros* sp. Although some elements such as *Hystrix rufosa*, occur elsewhere only in early and middle Villafranchian deposits (Etouairés, Villaroya), the remainder of the fauna is typically late Villafranchian in nature.

Some of the important later Pleistocene deposits have recently been reviewed in detail and their correlations discussed. Among these are: Předmostí (Musil, 1968), Voigtstedt in Thüringia (Kretzoi, 1965a and b; Janossy, 1965a and b; Kahlke, 1965a, b, and c; Thenius, 1965), Taubach (Kahlke, 1961), Süssenborn (Flerov, 1969; Kahlke, 1969a, b, c, d, e; Guenther, 1969; Fejfar, 1969).

Although some authors differ in their opinions on the detailed correlation of the various European Pleistocene faunas, the general sequence, with minor variation, has been accepted by most workers. A far greater problem is the establishment of intercontinental correlation with Africa and Asia, especially with the former.

Recently available radiometric age determinations indicate an earlier age for Villafranchian deposits than had generally been believed. These new data permit a tentative intercontinental correlation that is not inconsistent with faunal evidence. The age of the early Villafranchian deposits of Etouairés in the Perrier Hills of France is inferred to be about 3.4 m.y. by geological correlation with dated pumice believed to belong to the same horizon (Savage and Curtis, 1970; Azzaroli, 1970). The correlative deposits at Villafranca d'Asti, Laiatico, Viallette and Bethlehem are presumably of similar antiquity. An equivalent age range in Africa would include the basal Usno Formation, members B and C of the Shungura Formation and the later Kaiso Beds in East Africa, and the Chiwondo beds in Malawi.

On strictly faunal grounds such a correlation seems reasonable. Although several European genera do not appear in Africa until later Shungura or Olduvai Bed I time (e.g., *Felis*, *Lutra*), others such as *Hystrix*, *Hyaena*, *Anancus*, *Mammuthus*, and *Sus* were present in Africa during the Pliocene. The mammoth at Laiatico is a very primitive stage of *Mammuthus meridionalis* and differs little from the Ichkeul *Mammuthus africanavus*. *Anancus* persists into the middle Villafranchian of Europe and is present at Aïn Hanech, but its last occurrence in East Africa is in the Ekora beds.

The elephant at Bethlehem is an early form of *Elephas planifrons* (Hooijer, 1958) and was probably derived from a stage of *E. ekorensis* comparable to that occurring in the early Pliocene of Kanapoi and Ekora, dated at about 4.0 million years.

A K/Ar of 2.5 m.y. for Rocca Neyra, also in the Perrier Hills, gives an age estimate for middle Villafranchian deposits of Europe (Savage and Curtis, 1970). Radiometric correlation with Africa would indicate an age comparable to Shungura Beds D and E. The faunas agree well with such a correlation but the temporal ranges of most of the genera in common to these two continents are great enough to make correlation on faunal evidence alone inconclusive. *Vulpes* and *Lutra*, known from St. Valliers in Europe, do not appear in Africa until Olduvai Bed I age. *Hystrix*, *Hyaena*, *Anancus*, *Mammuthus*, *Hipparion*, *Canis*, and *Felis* are common to both continents during this period, the last two appearing for the first time in Africa. In Europe, *Equus* is first recorded at Pardines, and in Africa it first appears for certain at Koobi Fora in beds of equivalent age (2.6 m.y.).

The elephant from the Praetiglian, often referred to "*Elephas planifrons*" in the past, is an early stage of *Mammuthus meridionalis*, not very advanced beyond the Laiatico Stage of the early Villafranchian. At St. Valliers and Chagny a more progressive stage occurs, but this is somewhat less advanced than the typical Val d'Arno population. A similar grade of *Mammuthus* is also known from Africa, represented by *M. africanavus* at Ouadi Derdemi, Chad.

Savage and Curtis give an age of 1.9 m.y. for the solifluction deposits of Mount Coupet in Velay, presumably containing the upper Villafranchian elements of the Mount Coupet fauna (*op. cit.*). Azzaroli (1970) cites an age of 1.3 m.y. for a basalt overlying the fossil-bearing sediments at Sainzelles, also in Velay, and considered to be late Villafranchian in age. The dates correlate in Africa with Olduvai Bed I and lower Bed II and with the Ileret Member of the Koobi Fora Formation.

Hipparion (s.l.) had by this time disappeared from Europe, but persisted throughout Africa into the middle Pleistocene. The last occurrence of the genus *Anancus* in both Europe (Mosbach I) and North Africa (Aïn Hanech) is recorded at this time. Typical *Mammuthus meridionalis* of the Montavarchi Stage is morphologically inseparable from the Aïn Hanech elephant, but in East Africa *Mammuthus* had been replaced completely by *Elephas recki*.

ASIA

Interpreting the sequence of Pleistocene faunas of Asia has been complicated by the recognition of two faunal assemblages, the so-called "Sivamalayan" and

"Sinomalayan" faunas, the latter replacing the former in Java during the last part of the epoch. As was once the case in Africa, the relative ages of the Javan faunas have been the subject of heated debate, primarily because of early hominid remains known from the area. Unlike Africa, however, radiometric age determinations are exceedingly rare. In spite of its faunal complexity, a relatively straightforward faunal succession in Asia can be reconstructed.

The close relationship between the faunas of Burma and India was first emphasized by Pilgrim (1910) for the Lower Irrawaddies and Dhok Pathan deposits. Later, Stamp (1922) extended this analogy to later deposits, correlating the Upper Irrawaddies with the Tatrot horizon. Colbert (1943) considered the Irrawaddy fauna to be the easternmost extension of the Siwalik or "Sivamalayan" fauna characteristic of India, Burma, and the earlier deposits of Java (Tjidoelang, Kali Glagah; von Koenigswald, 1934, 1935, 1939). All of these are Pliocene to early Pleistocene in age.

By middle Pleistocene times a fauna of distinctly different character and one typical of southern China (Yenchingkou, Hoshangtung) extended southward into Burma (Mogok) and Java (Djetis, Trinil), with a characteristic assemblage consisting of *Macaca, Hylobates, Pongo, Ursus, Crocuta, Tapirus, Ailuropoda,* and *Stegodon* (Colbert and Hooijer, 1953; Hooijer, 1952). The contemporaneous faunas of northern China (e.g., Choukoutien) differed from the southern faunas in possessing a greater percentage of European elements. These faunas are roughly correlative with the Boulder conglomerate and Narbadda Valley faunas of India.

Pilgrim (1938) considered the faunas of the upper Siwalik Tatrot horizon and that of the Pinjor to be quite distinct, the former equivalent to the European Astian and the latter to the Villafranchian stage. Many authors (e.g. Colbert, 1943) have questioned such a degree of distinction between these assemblages. A recent study by Kahn (unpublished Ph.D. thesis in the Geological Department, Panjab University) demonstrates the occurrence of typically Pinjor species in the Tatrot horizon, thus confirming Colbert's earlier suggestion. These include *Stegodon insignis, Elephas planifrons* (also see Hooijer, 1955), *Sus.* sp., *S.* cf. *hysudricus, Camelus punjabiensis, Sivatherium giganteum, Leptobos falconeri, Hemibos triquetricornis,* and *Sivicobus palaeindicus.* However, the Pinjor is still distinguished by the appearance of taxa indicative of a somewhat later age; these include *Elephas hysudricus, Equus, Bos, Rhinoceros, Lutra, Mellivora,* and *Pongo.*

The fauna of the Ma Kai Valley in northern Yunnan is of the Siwalik type. The presence of *Rhinoceros, Bos,* and *Equus* suggests an age later than that of the Tatrot. The horse, *Equus yunnanensis,* is close to *E. sivalensis* of the Pinjor (Colbert, 1940). The remainder of the fauna is poorly known, but correlation with the Pinjor seems reasonable.

In northern China the faunas of Nihowan, Yüshe, and locality 18 near Peking contain faunas of late Villafranchian aspect. They are post-Pinjor in age, but are probably older than Choukoutien 1. *Bos* is absent, *Equus* is represented as *E. sanmeniensis,* and the elephant is close to *Elephas namadicus. Hipparion,* which does not survive beyond the Tatrot of India or the middle Villafranchian of Europe (Pardines, Rocca Neyra), is still present at Nihowan. The bear at locality 18 is close to *Ursus etruscus* of the European Villafranchian.

Much controversy has surrounded the age of the Pleistocene faunas of Java (Hooijer, 1952, 1956a, 1964; von Koenigswald, 1950, 1955, 1962). Since much of the debate has involved the relative position of these beds within the Pleistocene and the position of the Pliocene/Pleistocene boundary, it is irreconcilable on any but arbitrary grounds. Hooijer has argued that the Djetis beds are post-Villafranchian and, therefore, "middle Pleistocene" in age, and later than the Pinjor of India. The presence of *Hylobates, Pongo, Tapirus, Viverra,* and *Bos* would seem to link the Djetis fauna with the Szechuan fissure faunas of southern China, although this does not rule out the possibility of its being earlier. The elephant from the Djetis has been shown to be *Elephas planifrons,* whereas that from Trinil is *Elephas* sp., *E.* aff. *namadicus.* The former is typical of the Pinjor, disappearing before deposition of the Boulder conglomerate. The latter species appears at Nihowan and in the middle Pleistocene faunas of the Narbadda Valley of Central India, Yenchingkou of China and the Mogok Cave deposits of Burma. *Felis tigris* is common to both of these Javan faunas and to those of Yenchingkou (Szechuan) and Hoshangtung (Dien and Chia, 1938), but is absent from the Pinjor. Likewise, *Ursus malayanus, Hyaena* sp., *H.* cf. *sinensis* and *Tapirus augustus* of the Javan faunas also occur in southern China of late Villafranchian equivalent. The presence of a bovid related to but more advanced than *Leptobos* of the Pinjor also suggests a post-Pinjor age for the Djetis fauna (Hooijer 1956b). A recent K/Ar age of 1.9 ± 0.4 m.y. has been reported by G. H. Curtis for a tuff several meters below the *Meganthropus* site in the Djetis beds near Modjokerto (Stross, 1971). In spite of this early date, however, the age of the Djetis fauna would appear to be somewhat younger.

Elephas namadicus makes its first appearance in Asia at Nihowan, Yenchingkou, Mogok, and the Narbadda Valley deposits. In Europe it is first recorded from Mosbach I and the Cromer Forest Beds. The species persists until nearly the close of the Pleistocene, at Taubach in Europe and at Sjara-osso-Gol in Inner Mongolia. It is also known from Trinil and Ngandong in Java.

The faunal sequence at Choukoutien has been presented in a series of papers by Pei (e.g., 1936, 1939, 1940) and by Teilhard de Chardin (1936, 1938). Localities 1 and 3 appear to be younger than the Narbadda Valley fauna, and locality 3 may be younger than locality 1. The presence of such forms as *Elephas namadicus, Trogontherium cuvieri Felis pardus, Ursus spalaeus,* and *Ursus arctos* provide a correlation with the European middle Pleistocene faunas of Holsteinian age.

The Sjara-osso-Gol fauna (Boule *et al.,* 1928) and that of the upper cave at Choukoutien with *E. namadicus, Equus przewalskii, Equus hemionis, Cervus elaphus, Bos primigenius,* and *Homo sapiens* indicate a late Pleistocene age comparable to that of Ehringsdorf, Předmosti, and Taubach in Europe.

NORTH AMERICA

The relative sequence of local faunas of the Pleistocene in North America has been reviewed recently in an unpublished report compiled by the Cenozoic Correlation Committee of the Society of Vertebrate Paleontology. The correlation of North American faunas outlined here relies heavily on that report.

Correlation with Eurasiatic faunas cannot be exact because of the small number of genera that occur in both continents. Only nine genera are common to the Rexroad fauna of Kansas and European faunas of similar age. By the later Blancan, the number had increased to twenty. A sequence of radiometric age determinations has provided an additional means of correlating this North American sequence with that of Europe. The results are consistent with faunal data and suggest that the Blancan land-mammal age of North America is a correlative of the European Villafranchian (Savage and Curtis, 1970).

In the same paper Savage and Curtis report an absolute age of 2.7 to 3.1 m.y. for the earliest glacial deposit in the Northern Hemisphere (Sierra Nevada). Hibbard *et al.* (1965) give 2.5 m.y. to 4 m.y. as the age of the earliest Blancan faunas which antedate the first glacial phases in North America. In a correlation of deep-sea cores using per cent abundance of the foraminifera *Globorotalia menardi*-complex and its direction of coiling, and using absolute dating, Ericson and Wollin (1968) set up an absolute time scale based on interpolation from magnetic events of known age (Olduvai normal, 1.9 m.y.; Jaramillo normal, 1.0 m.y.). They suggest an absolute time scale for North American glacial events with the Nebraskan at about 1.7–1.9 m.y., the Kansan at 1.0–1.4 m.y., the Illinoian at 0.4–0.5 m.y., and the Wisconsin less than 0.15 m.y. This approach must be viewed with caution, however, especially in view of the seemingly greater frequency of magnetic events than has previously been believed (Cox, 1969). Nevertheless, from the scant faunal evidence and from available glacial and radiometric data, the above correlation is not too unreasonable, and is close to that presented in figure 12. More recent analysis of deep-sea cores in the eastern equatorial Atlantic suggests somewhat more recent ages for the earlier cold pulses (Ruddiman, 1971). Climatic deterioration was marked by short but cold phases beginning at 1.3 m.y. A second, longer series of cold intervals occurred from 900,000 to 775,000 years B.P. and again at 600,000 to 425,000 years B.P. It is probably too early to attempt to draw any firm conclusions regarding the correlation of these events with continental sequences. Yet the vertebrate evidence would seem to favor a somewhat older age for the earliest cold phases, at least in those few areas where such faunas and absolute dating have been correlated.

VI. ORIGIN AND PHYLOGENY

The establishment of probable phyletic relationships is essential to the understanding of evolutionary phenomena based on the fossil record. In the absence of a stratigraphically confirmed phyletic continuum, time-correlated morphological (and presumably adaptive) change must be inferred from level of structural complexity, or grades. These may have little reality in terms of continuously adapted populations undergoing progressive evolution, and, as will be clear from the discussion below, the study of grades alone often can be misleading. Studies of evolutionary rates are impossible without well-documented lineages with adequate time control.

Because of the lack of direct observation in paleontology, segments along a lineage are phylogenetic by inference only. In a few cases, gaps between successive segments are so small that inference approaches certainty, although it can never reach it. A phylogeny is an ancestral-descendant succession of populations genetically continuous through time. It is, therefore, an objective biological relationship recognizable through morphological, ecological, and stratigraphic (temporal) analysis of fossil data. As Simpson (1959) has emphasized, populations along the lineage do not form a phylogeny *because* they are morphologically similar; on the contrary, they bear certain similarities because of the phyletic relationships between them. It is, therefore, possible to recognize phyletic lines, or at least to approximate them, if the appropriate fossil material is available. If this were not so, fossils would hardly be worth studying at all.

The establishment of a meaningful phylogeny that accurately reflects past evolutionary events is a difficult task at best, and often only a rough outline can be reconstructed. When dealing with taxa above the species level it is often difficult to deduce correctly the actual ancestral-descendant sequence as opposed to successive, but discontinuous, non-ancestral offshoot taxa that form a series of grades. When dealing with taxa on the species level or below, the problems are less acute, although still present. In most cases, fossil material does not allow the degree of resolution necessary to trace actual ancestral populations. Nevertheless, in a number of Pleistocene groups, including the Ele-

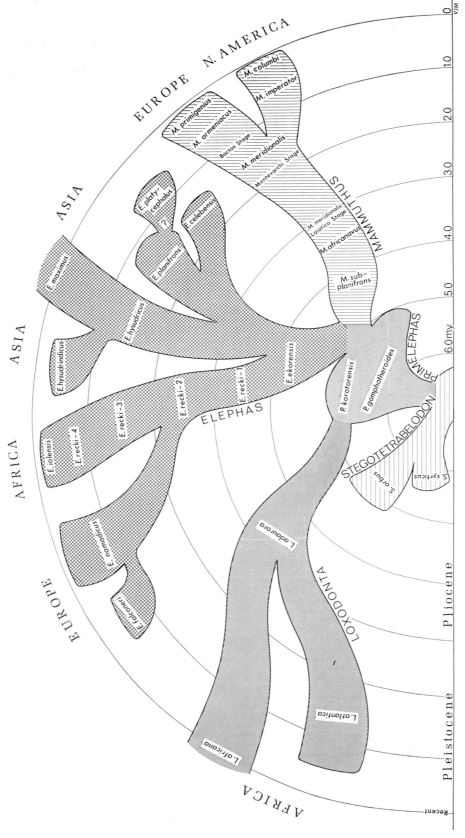

FIG. 13. Proposed phylogenetic relationships between all species of the family Elephantidae recognized as valid in the present study.

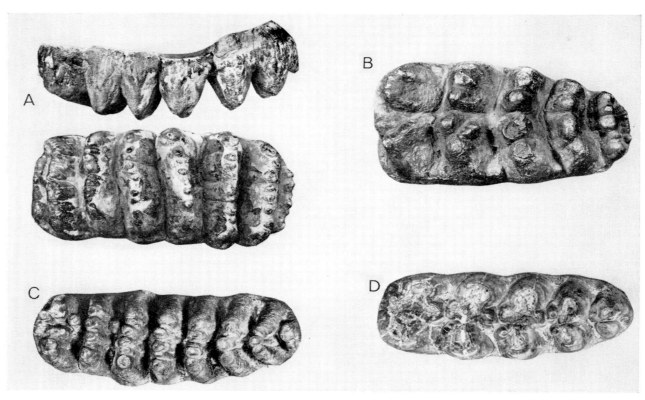

Fig. 14. Comparison of molar structure in a stegodont, a stegolophodont, an early elephant, and a gomphothere. Lateral and crown view. A: *Stegodon insignis*. B: *Stegolophodon cautleyi*. C: *Stegotetrabelodon orbus*. D: *Gomphotherium* cf. *angustidens*.

phantidae, it is now possible to reconstruct a phyletic sequence of species and stages that appears to approximate the actual evolutionary succession.

Numerous phylogenies of the Elephantidae have been proposed in the past and in figure 2 we saw four of the more divergent of these views. The phylogeny suggested in the present study (fig. 13) differs from most others on a number of minor, and in some cases major, points; it is closest to that presented by Aguirre (1969).

One of the problems in interpreting the structural and functional modifications of the elephant dentition arose from the false notion that elephants were derived from stegodonts, an idea held by most students of the group until very recently. Starting from the specialized structure of the stegodont molar (fig. 14A) with its compressed transverse valleys, complete loss of intravalley columns and median folds, and consolidated plates seemingly composed of a series of pillars fused side by side, it was impossible to understand the evolution of the elephant molar, which is basically so different. Aguirre (1969) and I (1970a) have suggested that the immediate ancestry of the Elephantidae must be sought among the gomphotheres of Africa, although Aguirre views the stegolophodonts as an intermediate group, which I do not. The teeth of *Stegolophodon* are already too specialized in the direction of *Stegodon* to have served adequately as the basal elephant type.

Starting with a gomphothere molar, however, the elephant dentition begins to make some sense.[15]

The gomphothere origin for the elephants became clear with recognition of the significance of *Stegotetrabelodon syrticus* Petrocchi from the late Miocene beds at Sahabi in Libya. Petrocchi (1941) considered his new genus to represent a specialized ("early Miocene") gomphothere, but it was later argued that this form is best considered a primitive elephant (Maglio, 1970b). Aguirre (1969) pointed out the elephantlike characters of the *Stegotetrabelodon* dentition, but did not discuss its phyletic relationships to the Elephantidae.

Stegotetrabelodon syrticus possesses a large pair of mandibular incisors, as in gomphotheres generally, but the symphysis is beginning to become reduced in size. The only mandible described for this genus (there are at least two additional undescribed specimens) has a backwardly sloping ramus and a prominent angle with a projecting flange for attachment of the medial pterygoid and superficial masseter muscles. These osteological features are very common in the Gomphotheriidae but absent in all later elephants. The skull is incom-

---

[15] While this paper was in press, proboscidean material was collected by W. W. Bishop from Ngorora near Lake Baringo, Kenya. This site is dated radiometrically at 9–11 m.y. The specimens are clearly intermediate between *Gomphotherium* and *Stegotetrabelodon*.

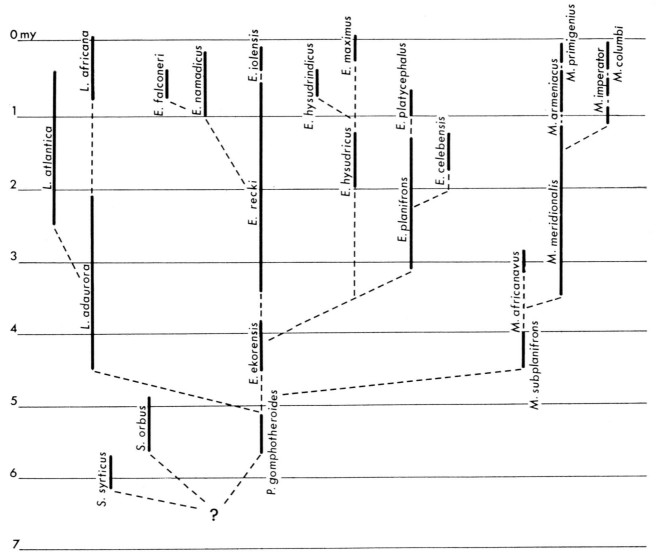

Fig. 15. Time distribution of the twenty-five species of elephants recognized as valid in the present review. The ranges represent the total time span covered by known records of occurrence.

pletely known (see fig. 7) but offers no problem in serving as a base for later elephantid evolution. It is in the dentition, however, that we find a clear relationship to the Elephantinae which cannot be found in any other proboscidean group.

The molar teeth exhibit a number of typically gomphothere characters modified into a basically elephant-type shearing battery. These include a low crown with very thick enamel that is unfolded and bears little external cement. The gomphothere pretrite and posttrite columns (see page 88 below) are here transversely elongated and subdivided into two pillars each. They meet in the midline to form a platelike structure with a median cleft separating the two columns for about one-half of their height. Strong intravalley columns arise out of the base of these plates; these correspond in structure and position to the "trefoil folds" of gomphothere molars and were almost certainly derived from them. Similar columns are found in all earlier species of elephants. The low number of plates, the V-shaped valleys between plates, the unfolded thick enamel and the wide spacing of the plates are features typical of early elephants and represent only slight modifications of the gomphothere molar pattern.

At present the genus *Stegotetrabelodon* is known from incomplete material in Miocene/Pliocene deposits, the oldest being associated with dated trachytes of approximately 7.0 m.y. age (W. W. Bishop, pers. comm.). It is the only proboscidean in the Kaperyon Beds (more than 5.0 m.y.). A new species from Chad demonstrates

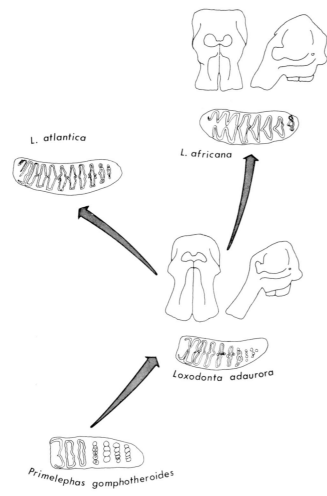

Fig. 16. Diagrammatic representation of the phyletic relationships between species of *Loxodonta* and *Primelephas gomphotheroides*. Molar and cranial morphologies are illustrated where known.

the group's existence in central Africa during this period. It is probable that the Stegotetrabelodontinae will prove to have been a dominant proboscidean group for at least the latter part of the Miocene in Africa. It appears to have been the basal group out of which all later elephants arose.

The more progressive forms of elephants have been placed in the subfamily Elephantinae. The earliest known species are still rather primitive, but show certain modifications which represent a further shift into the elephant adaptive zone. *Primelephas gomphotheroides* is the best known of these species; it occurs in Lothagam 1 and in the lower Kaiso Formation at Nyawiega. The molars show structural advances over the condition found in *Stegotetrabelodon*. As in all later elephants, the plates are more consolidated and the median cleft is lacking, except on the first plate. The total number of plates is greater by one or two and the individual pillars are less well defined. The valleys between plates are more open at the base, especially in M1 and M2. As a result, the plates are less triangular in cross section, being more parallel-sided, although less so than in later elephants. The mandible has a very short symphysis but retains a pair of small incisors. The latter are externally ribbed and transversely compressed as in *Stegotetrabelodon*.

From the morphological evidence there would seem to be little doubt that *Primelephas* was derived from the Stegotetrabelodontinae in Africa sometime during the latest Miocene. Its features, although more strikingly elephantine in structure, are merely further developments of the trends already well underway in that group. *Primelephas gomphotheroides* occurs with *S. orbus* in Lothagam 1 and is therefore unlikely to have been derived from that species. The large size and very primitive nature of the Sahabi species (*S. syrticus*) would seem to preclude it from direct ancestry. A third species from Chad already had undergone reduction in the mandibular incisors and may have been closer to the base of the Elephantinae.

By Kanapoi time in Africa (4.0 m.y.), three lineages of elephantine proboscideans had become established (fig. 15), each distinctive in many important cranial characters, but broadly similar in dental morphology. The teeth in all groups show still further advances over the *Primelephas* grade, being higher crowned and having a greater number of plates. The latter are more closely spaced and their enamel covering is thinner. Transverse valleys are more broadly open and U-shaped at the base.

One of these lines includes the living African elephant and its congeners (fig. 16). The earliest recognizable stage of *Loxodonta* is *L. adaurora* known from Kanapoi, Lothagam 3, the Mursi Formation, etc. The structure of the skull and jaw leave no doubt that this form was on or very near to the direct line of ancestry leading to the living species. Except for a relatively few distinctive traits, *L. adaurora* already possessed all of the modifications characteristic of the genus. In the dentition the posterior columns are only partially fused into the plates and form large loops with wear, but the median expansion of the plates so typical of later species of *Loxodonta* is not yet evident. The vestigial mandibular incisive root chambers, the short, but prominent symphysis, the long corpus and general molar structure point to *Primelephas* as the ancestral group for *Loxodonta*. Stratigraphically, the evidence also favors such a relationship.

The several specimens of *L. adaurora* known from deposits at Koobi Fora and in member G of the Shungura Formation do not differ significantly from the Kanapoi material. Although the species is known at this level only from isolated teeth, it appears to have remained relatively unchanged during the previous two million years, at least with respect to its dentition.

From this point in the early Pleistocene the genus became rare, disappearing completely in the upper Shungura equivalent in East Africa and reappearing, as *Loxodonta atlantica,* in middle Pleistocene deposits of North and South Africa, at Ternifine and Elandsfontein respectively. *L. atlantica* is represented by numerous isolated teeth and a fragmentary skull. The molars are more progressive than in *L. adaurora* in having a greater number of plates and higher crowns. Unlike other species of the genus, the enamel is highly folded and the median loops are enlarged and often distinctively bifurcated. Irregular smaller loops occur around the worn, lozenge-shaped enamel figures. The skull seems to be typically *Loxodonta* in form, but the occipital condyles are enormously enlarged. The tusks are relatively small and gently curved as in all other species of *Loxodonta*. The *Loxodonta* affinities of this species seem fairly certain and *L. adaurora* could easily have been ancestral to it. The first appearance of *L. atlantica* is in Bed E of the Shungura Formation and it was, therefore, contemporaneous with the latest stages of *L. adaurora*. It is probable that *L. atlantica* originated during the time represented by the earlier Shungura deposits.

The living African elephant, *Loxodonta africana,* is known in the fossil state in late early Pleistocene to sub-Recent deposits. Its teeth are in many ways less progressive than those of *L. atlantica*. The enamel is not folded, the crown is lower, and the number of plates is not as great as in that species. The median enamel loops are simple, as in *L. adaurora*. Like *L. atlantica,* the premaxillaries and tusks are reduced in size and the mandible is shorter than in the Kanapoi species. *Loxodonta atlantica* does not appear to have been on the main line leading to the living species, but rather to be a somewhat more specialized collateral line. *L. africana* is best derived from *L. adaurora* from which it differs in remarkably few ways.

The second phyletic group already established in the early Pliocene beds at Kanapoi, Ekora, and Kubi Algi is that culminating in the living Asiatic elephant and other species of the genus *Elephas* (fig. 17). *Elephas ekorensis* is the earliest known species and is associated with *L. adaurora* at Ekora and Kubi Algi. The teeth are already somewhat more progressive than in *Loxodonta,* but not greatly so. The plates are a little narrower and the enamel is thinner. The skull is very distinct and clearly of the *Elephas* type. The anteroposterior compression, expanded parietals, swollen occipitals, strong parietal ridges, etc. that characterize all species of *Elephas* are already present. The tusk alveoli are parallel and already rather small. The Kubi Algi specimens are less advanced in all of these features, as would be expected from their earlier age.

Elephas recki is the best known species of *Elephas* from Africa; it can be traced through a progressive series of stages from an early form in the middle Pliocene of Kikagati to a rather progressive form in the middle Pleistocene of Bed IV at Olduvai Gorge. These stages pass almost imperceptibly into each other, as can be seen in the successive populations known from Kaiso, Omo, Laetolil, Koobi Fora and Ileret, Ouadi Derdemi, Olduvai, Kanjera, and Olorgesailie, among others. The molar teeth show progressive increase in the number of plates, their relative height and spacing, reduction in enamel thickness, increased folding of enamel and loss of the median loop on the wear figures. The skull shows all of the modifications already begun in *E. ekorensis,* but here they are developed even further. On stratigraphic and faunal evidence the earliest stage of *E. recki* (Kikagati) would have been later than the latest record of *E. ekorensis*. At present, a direct phyletic relationship between *E. ekorensis* and *E. recki* seems certain.

In the late Pleistocene of Africa a more progressive elephant appears which I retain as a distinct species, *Elephas iolensis,* only as a matter of convenience. Although as a group, material referred to *E. iolensis* is distinct from that of *E. recki,* some intermediate specimens are known and *E. iolensis* seems to represent a very progressive, terminal stage in the *E. recki* specific lineage (see fig. 17). It is only known from teeth, but these show a continuation of the same trends begun in *E. ekorensis* and developed through the successive stages of *E. recki*. Lacking evidence to the contrary, there is no reason at present not to consider *E. iolensis* as the direct descendant of *E. recki*.

In Europe and Asia, *Elephas namadicus* makes its first appearance during the Cromerian and persists into the latest Pleistocene. It was formerly believed to have evolved from *Mammuthus meridionalis* as a collateral branch with *M. armeniacus*. However, the former species is still present at Bacton where it possesses none of the *Elephas* features in the skull and dentition so typical of *E. namadicus*. The expanded parietals and occipitals, sharp parietal ridges, nearly straight, untwisted tusks, the very deep median-parietal depression and the transversely concave forehead set this species apart from the *Mammuthus* phyletic group. *E. namadicus* shows too many specializations in skull and dentition to have arisen in the short time available if *meridionalis* were its immediate ancestor. If it were derived from a very early stage of *meridionalis,* then we would expect to find some trace of it in Europe during the Villafranchian; we do not.

The structure of the molar teeth in *Elephas namadicus* is specialized in having a reduced median fold and a slightly expanded, sharply pointed vertical ridge on the plate face. When worn, this gives a vague lozenge-shape to the plate. This is, however, very variable. The nature of this plate structure is not like that of *Loxodonta* in which the entire median portion of the plate is expanded. The skull also rules out any pos-

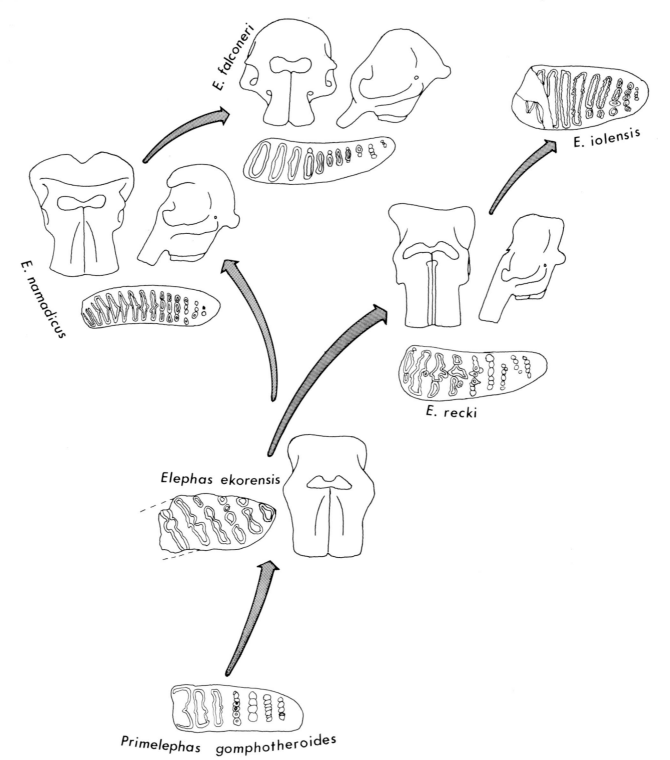

Fig. 17. Diagrammatic representation of the phyletic relationships between the five species of the African and European branch of the *Elephas* complex, and *Primelephas gomphotheroides*. Molar and cranial morphologies are illustrated where known.

sibility of close relationship between *E. namadicus* and the *Loxodonta* phylum.

Except for the above mentioned molar specialization, the dentition of *E. namadicus* closely approaches that of the *E. ekorensis-E. recki* lineage, but the molars are narrower and more progressive in the number and spacing of plates and in the crown height. The skull shows a close similarity to that of the *E. recki* line. The forward expansion of the fronto-parietal region occurs in both, but in *E. namadicus* it is much more developed, projecting forward as a prominent crest. The tusk sockets are widely spaced as in *ekorensis*, but they flare distally in *namadicus*. The temporal region is broader than in *E. recki* and the naris is less elongated at the sides.

Although the time and place of origin of *Elephas namadicus* is still uncertain, it seems to have been derived from a form close to *E. ekorensis* or perhaps an early stage of *E. recki*. Its sudden appearance in Europe and in Asia during the Cromerian equivalent suggests immigration from elsewhere and an origin perhaps as early as the beginning of the Pleistocene. Just where the species evolved is at present a mystery as transitional forms are unknown. The Middle East and central Asia seem good possibilities in this regard. Earlier Pleistocene deposits are rare in these areas so that this suggestion is pure speculation at this time. By the middle Pleistocene the species was abundantly present in Asia and several fragments of elephant teeth from Ubeidiya in Israel (Haas, 1963, 1966), though too incomplete for identification, may belong to an early stage of *E. namadicus*. The Ubeidiya fauna appears to be middle to late Villafranchian in age [16] and additional elephant material from this locality would be most welcome. For the present, then, we may suggest an origin of *Elephas namadicus* from the African *E. recki* lineage, probably during the early Pleistocene.

The pygmy elephant of Sicily, *Elephas falconeri*, is distinct from *E. namadicus* in a number of important features, but these are all almost certainly related to dwarfing. The skull lacks the frontal crest and has all the juvenile characters, such as large orbits, small tusks and tusk sockets, rounded cranium and weak basi-cranial flexure that are seen in young specimens of other elephants of normal size (see fig. 18). The teeth have relatively fewer plates with thicker, less folded enamel than in *E. namadicus*. As will be discussed below, these traits also are functionally correlated with dwarfing. A number of specimens from Sicily and other Mediterranean islands and intermediate in structure between *E. namadicus* and *E. falconeri* show less drastic effects of paedomorphosis and clearly demonstrate the *namadicus* ancestry of these pygmy elephants (fig. 17).

In a recent publication on elephant phylogeny, Aguirre (1969) proposed to include "*Elephas mnaidriensis*" of Sicily in the genus *Loxodonta* because of the

[16] The cultural artifacts from this site have been identified as Israeli Variant of Oldowan I and II, and Israeli Variant of Abbevillian (Tobias, 1966), and would in part suggest a somewhat later age.

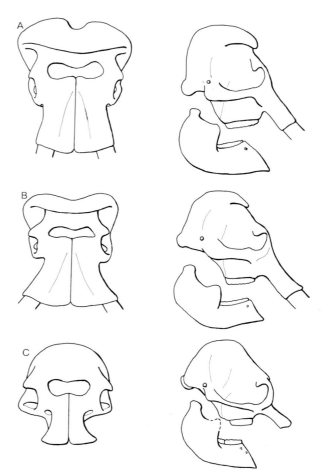

FIG. 18. Diagrammatic representation of skull morphology for three species of *Elephas*, showing the effects of progressive dwarfing. A: *E. namadicus*, × 1/24. B: *E. mnaidriensis*, × 1/12. C: *E. falconeri*, × 1/9.

morphology and thick enamel covering of its molar teeth, despite the fact that it is "strikingly similar" in skull morphology to *E. namadicus*. There is no question that the skulls of these two forms are similar, so much so that it is difficult to imagine an independent *Loxodonta* origin for this species. The differences in the molar teeth seem certainly related to dwarfing and may be compensatory modifications for the maintenance of molar efficiency (Maglio, 1972b; and page 94 below). On the available evidence, I believe we must regard all the Mediterranean pygmy elephants as derivatives of the *E. namadicus* group.

The earliest elephant known from Asia is *Elephas planifrons*, often considered the direct ancestor of *E. hysudricus* and *E. maximus*. As has been shown (Maglio, 1970b), the skull of *E. planifrons* is remarkably close to that of *E. ekorensis* in most details, but differs in several important ways (see fig. 11). None of these would preclude a close relationship between the two species. The dentition of *E. planifrons* is slightly

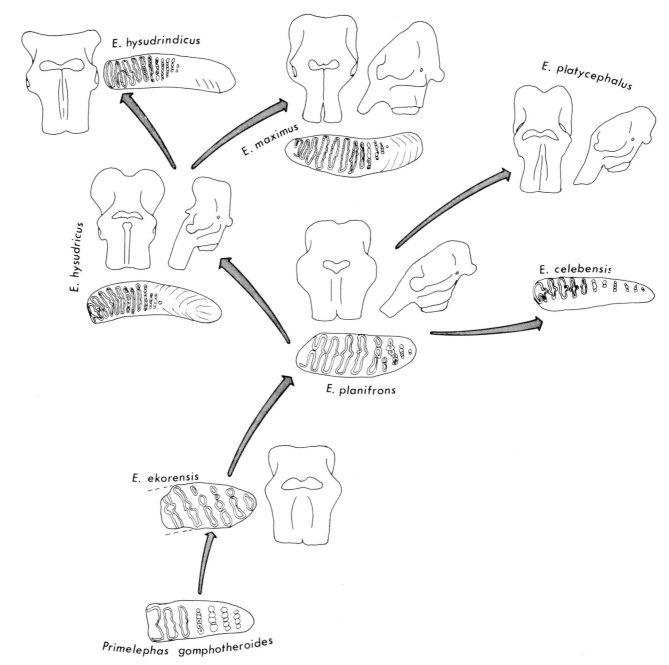

Fig. 19. Diagrammatic representation of the phyletic relationships between seven species of the Asiatic branch of the *Elephas* complex and *Primelephas gomphotheroides*. Molar and cranial morphologies are illustrated where known.

more primitive than that of typical *E. ekorensis* from Ekora, but is close to the Kubi Algi stage of that species.

The Bethlehem elephant represents one of the more primitive populations of *E. planifrons* known and is very close to the earliest form of *E. ekorensis*. Bethlehem is geographically intermediate between the ranges of *E. ekorensis* and *E. planifrons* and stratigraphically it would seem to represent a period between the latest occurrence of *E. ekorensis* at Ekora and the first appearance of *E. planifrons* in the Tatrot horizon of the Siwalik Hills. Before the discovery of *E. ekorensis*, the ancestry of *E. planifrons* was open to question. Discovery of *E. ekorensis* in Africa provided a morphologically and stratigraphically suitable ancestor. In view of the Bethlehem material, however, a somewhat earlier stage of this species in Africa, one close to the stage represented at Kubi Algi, may have been the actual source for Asiatic *Elephas*.

The previously supposed relationship between *E. planifrons* and *E. hysudricus* offers some problems on morphological grounds. The skull of *E. planifrons* is relatively less specialized in having a broad, flat forehead, widely spaced orbits and strongly reduced, forwardly directed tusks. The narial aperture is small, high on the head and upturned at the sides. The dentition, on the other hand, is variable in *planifrons* and this probably reflects evolution of the species through time. On strictly anatomical grounds this species could not have given rise to *E. hysudricus* (fig. 13). The skull of *E. hysudricus* differs strikingly from that of *E. planifrons*, and more closely resembles that of *E. recki*. The tusks are still of modest size, the forehead is even narrower than in *recki*, and the naris is large and strongly downturned at the sides as in that species. The frontal expansion is similar in kind, but even greater in degree than that of *E. recki*.

From stratigraphic evidence it is not possible for *E. planifrons* to have been simply an early stage of the *E. hysudricus* lineage; both species are present in the Pinjor horizon of the Siwaliks (both may have also been present in the Tatrot) and *E. planifrons* seems to have persisted into the Djetis Beds of Java. The two species thus seem to have overlapped completely. The latter species was a persistently primitive collateral line of the Asiatic branch of the *Elephas* complex. It seems probable that both *Elephas hysudricus* and *E. planifrons* had a common ancestry from the *ekorensis-recki* complex, with *hysudricus* closely paralleling *E. recki*, and *planifrons* retaining more primitive features.

The stratigraphic relationships of the deposits bearing *Elephas celebensis* at Beru and Sompoh, Celebes are unknown, but Hooijer (1949) believes them to be early Pleistocene in age. Although only known from teeth and jaws, *E. celebensis* is so close to *E. planifrons* that a direct phyletic relationship seems highly probable. The presence of tusks in the lower jaw led Hooijer to propose an independent origin for *E. celebensis*. However, vestigal incisive chambers persist in *E. planifrons* and the redevelopment of tusks in *E. celebensis* evidently resulted from dwarfing.

The temporal distribution of *Elephas hysudricus* covers only a short period, from the Pinjor and Upper Irrawaddy Beds to the lower Karewa Beds, and affords little opportunity to trace successive stages through any significant period of time. As presently known, this species appears to have been on the phyletic line leading to the living Asiatic elephant. The structure of the skull in the latter has a less extreme parietal expansion and the temporal region is proportionately less constricted. Aside from this, the two species are very close.

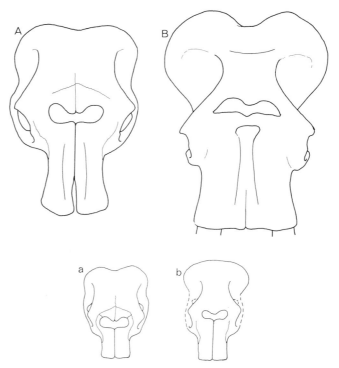

Fig. 20. Ontogenetic changes in skull morphology for two species of *Elephas*. A: *E. maximus;* B: *E. hysudricus*. A and B, adults. a and b, juvenile. Not to scale.

The premaxillary region is more reduced in *E. maximus* but this is correlated with the reduction of the tusks.

The absolute size of *E. hysudricus* is greater than *E. maximus*. An examination of the ontogenetic development in the latter, which is revealed by six skulls ranging in age from young juvenile to adult, shows that **the marked temporal constriction** and parietal expansion, plus other cranial modifications associated with these changes, are late growth developments (fig. 20). It is therefore not difficult to derive the smaller *E. maximus* from *E. hysudricus* by retardation of ontogenetic development, which would result in a less specialized-looking skull. In the dentition, *E. maximus* continues the trends seen in *E. hysudricus* but is more progressive in every way.

Another Asiatic species of the genus *Elephas* is *E. hysudrindicus* from Java, originally believed to be transitional between *E. hysudricus* and *E. maximus*. Indeed, the molar teeth are exactly what would be expected of such an intermediate form, as is the age of the beds at Tinggang, Kedoeng, and Ngandong which is middle to late Pleistocene. In the skull, however, *E. hysudrindicus* exhibits a remarkable degree of convergence toward the skull of the earlier *E. recki*. The parietals are flattened dorsally and expanded both forward and laterally. The skull is more gently compressed in the facial plane than in either *E. hysudricus* or *maximus*. *Elephas hysudrindicus* is certainly part of

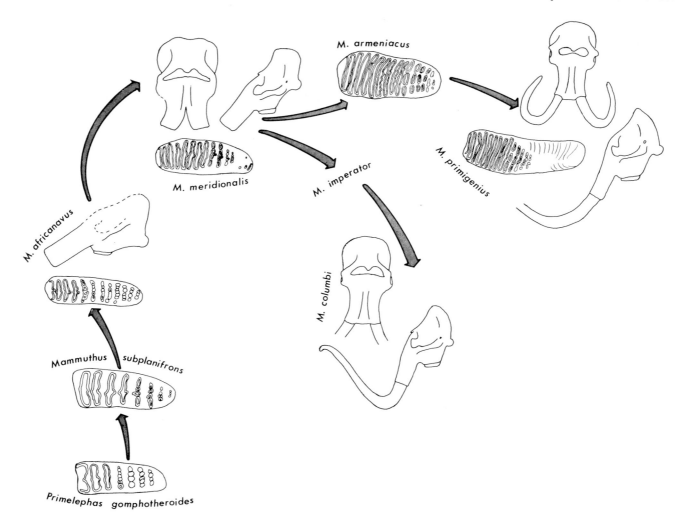

FIG. 21. Diagrammatic representation of the phyletic relationships between species of *Mammuthus* and *Primelephas gomphotheroides*. Molar and cranial morphologies are illustrated where known.

the *Elephas* phyletic group and could have been derived from an earlier stage of the *E. hysudricus* specific lineage. It does not appear to have been on the line leading to the Recent *E. maximus*.

The third phyletic group originating in Africa during the late Pliocene was that of the mammoths (fig. 21). All members of this group are referred to the genus *Mammuthus*. The earliest known form is *M. subplanifrons* known from the Vaal River gravels, Langebaanweg, Kanam West, the Chemeron Formation and from the Chiwondo beds. In the Chemeron and Chiwondo beds teeth of this type appear to have been contemporaneous with *Loxodonta adaurora*.

It was recently pointed out (Maglio and Hendey, 1970) that the "species" *M. subplanifrons* is probably a heterogeneous grouping of primitive-looking teeth. The specimen from Virginia, O.F.S. is associated with a typically *Mammuthus*-like maxillary tusk. A mandible from horizon I at Langebaanweg has been shown to differ from *Loxodonta* in a number of ways and to be referable either to *Mammuthus* or to *Elephas*. Thus, at least part of what is now called *M. subplanifrons* is probably ancestral to the later mammoths, but some referred specimens from various localities may actually belong to early stages of *Loxodonta* or *Elephas*. It is primarily because of the clear *Mammuthus* affinities of the Virginia specimen, which agrees closely with the type, that this taxon is considered to be a part of the mammoth lineage. When better known, some of the East African material now included here may have to be transferred elsewhere.

Mammuthus subplanifrons, as currently defined, has molar teeth that are very primitive in structure, being low-crowned and thick-plated with few plates. Yet, it is considerably advanced over the *Primelephas* stage in that the plates are parallel-faced, the crown is higher and the transverse valleys are broadly open at the base. As already mentioned, the tusks were spirally twisted as in later mammoths, and presumably the skull was also modified, but evidence is lacking. The mandible as seen in the Langebaanweg specimen bears a pair of incisive chambers, as in early stages of *Loxodonta* and *Elephas,* but there were no external tusks in the jaw. Evidence relating to the origin of this form is not yet available, but its appearance in the early Pliocene together with *Loxodonta adaurora* and *Elephas ekorensis* suggests that all three forms were part of a single radiation. *Primelephas* was generalized enough structurally to have been the ancestral group of *Mammuthus* as well as of the other elephant groups.

A later species of mammoth, known for certain only in northern Africa, is *M. africanavus*. The skull and tusk morphology leaves little room for doubt as to the relationship of this species to the mammoths. The teeth are at about the same evolutionary stage as those of *L. adaurora*. Advances in dental structure were correlated with functional adaptations that were, by and large, similar in all lines of Pleistocene elephants in their early stages. The teeth of *M. africanavus* are more progressive than those of *M. subplanifrons,* but do not differ in basic structure. North African localities with *M. africanavus* seem best to correlate with European Villafranchian. The presence of *Equus* would suggest a middle Villafranchian age, but the temporal distribution of this genus in Europe and Africa is still uncertain. It seems unlikely that Ichkeul could be much younger than early to middle Villafranchian, by which time a very early stage of *Mammuthus meridionalis* was already present in Europe. On morphological grounds *M. meridionalis* could only have been derived from an early form of *M. africanavus*. The Laiatico Stage of *meridionalis* represents the beginnings of this lineage in Europe and is structurally very close to *africanavus*. The latter probably represents a persistent African population.

The presence of an elephant at Aïn Hanech, Algeria, that is structurally equivalent to late Villafranchian stages of *M. meridionalis* in Europe suggests either expansion of the latter species into North Africa or the persistence of *M. africanavus* there with parallel development toward the *meridionalis* grade of molar structure. Aïn Hanech is roughly equivalent in age to late Villafranchian deposits of southern Europe. The geographic relationship of these two areas makes the occurrence of a single species on both sides of the Mediterranean seem less unlikely. Until this North African form is better known, it is probably best to retain it under the name *M. meridionalis* as originally described.

The evolution of *M. meridionalis* in Europe can be traced through the Villafranchian to a progressive form in the Bacton Forest Beds. The species passes both morphologically and stratigraphically into *M. armeniacus* and this into *M. primigenius*. The last two "species" certainly represent arbitrary segments of a continuous lineage along which progressive molar and cranial changes occur through the middle and late Pleistocene.

The North American mammoths form an independent branch derived from a form of *M. meridionalis* that must have entered the continent late in early Pleistocene times. The successional species, *M. imperator* and *M. columbi,* paralleled the *M. armeniacus-M. primigenius* line in both cranial and molar structure. The lineage differs from the European branch in having broader molars, a more massive jaw, and certain cranial distinctions, but it certainly represents a collateral branch of the genus in North America.

VII. EVOLUTIONARY TRENDS
EVOLUTION AND FUNCTION OF THE ELEPHANT DENTITION

As discussed earlier, the true elephants originated in Africa from a gomphotheriid ancestor. The structural modifications that occurred in the dentition were elaborations of the basic structure of the gomphothere molar. Progressive adaptive shifts in the function of the masticatory apparatus, probably in response to shifts in habitat utilization during the Pliocene, resulted in the unique, highly efficient molar of the elephant group. Once established, the superbly adapted skull and dental complex underwent relatively minor (in a structural sense) alterations in response to selective pressures. To understand the evolution of the elephant dentition, we must first consider the structure of the gomphothere dentition.

The structure of the dentition in *Gomphotherium* from Miocene deposits of Africa and Europe provides the foundation upon which later specializations were built. Except for the first premolar, which was already lost in the earliest known proboscideans, *Gomphotherium* retained functional premolars that replaced deciduous predecessors. Of these, dm2 was least molarized, consisting of only two pointed cones separated by a transverse valley and flanked medially by one or more accessory cusps rising out of the cingulum. Dm3 and dm4 were strongly molarized with two pairs of cones in the former and three in the latter. A deep longitudinal cleft separated the lingual member of each pair from the buccal one. A strong accessory column arose from the valley floor, more closely associated in the upper molars with the lingual cone behind it, and in the lower molars with the buccal cone in front of it. The premolars were roughly square in outline with four cones separated by a broad transverse valley and a narrow anteroposterior cleft. Tiny accessory ridges were often present.

The permanent molars were complex structures consisting of three (M1–M2) to four (M3) pairs of cones, each pair separated by a very broad, V-shaped transverse valley (fig. 22). A deep, but compressed median

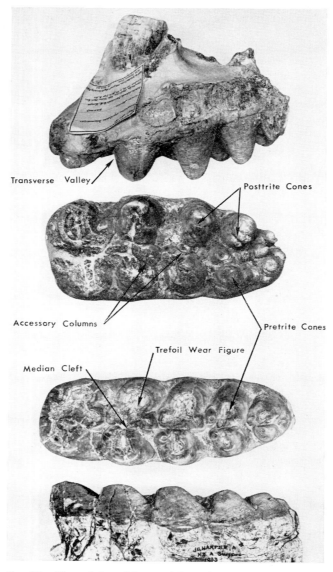

Fig. 22. Upper left and lower right molars of *Gomphotherium angustidens* illustrating the essential features of the gomphothere dentition.

cleft separated the lingual and buccal cones down to the base of the crown. Each cone was somewhat elongated transversely at the apex but was rounded toward the base. The lingual cones on the lower molars and the buccal cones on the upper ones were simple in structure with a tendency toward subdivision. The apical divisions often persisted partially down the anterior and posterior faces of the cones as weak grooves. On M3 there were up to three divisions. Accessory columns in the valleys were associated with the buccal cones on the lower molars and with the lingual cones on the upper ones; these were situated near the median cleft and fused with the associated cones. With wear, these cones and fused columns formed a three-lobed, or trefoil, pattern. The median lobes of successive cones came into contact at the midline. Additional minor columns and folds were often present. The enamel was very thick, extremely rugose and not folded.

The succession of the teeth in the jaw had become modified from the original mammalian type because of the great length of the molars, even in these early forms. The milk molars erupted successively and were present together in juvenile individuals. Vertical replacement by the third and fourth premolars followed, but these were shed as M2 and M3 came into place. M1 was usually shed before eruption of M3 so that in an adult only the last two molars remained. M2 was shed in very old individuals leaving only M3.

Functionally, the gomphothere molar is a grinding-shearing structure that apparently described a complex pattern of movement during mastication. From scratch marks on specimens in various stages of wear it is clear that the mandible was not confined to a primarily fore and aft motion as in elephants, but had an added circular motion with strong lateral components. The lower teeth occluded with the uppers while the jaw was in a retracted position. Then moving upward, forward and medially (probably inward on the occluding side), the lower cones described a circular motion on the uppers.

As wear proceeded, the buccal cones on the lower molars and the lingual cones on the upper ones—the cones with accessory columns and a trefoil type of wear—invariably were worn more deeply than their neighbors, on the opposite sides. This resulted from the manner in which the teeth rotated into position—the lowers rotated outward as they advanced forward. These first and more deeply worn cones are referred to as the *pretrite* cones, their neighbors as the *posttrite* cones. Because all cones are considerably broader at their base, the pretrite cones are broader on the worn occlusal surface than the corresponding posttrite ones. Consequently, the pretrite cones soon come to consist of a trefoil-shaped enamel loop enclosing a broad basin of dentine. With the forward and circular movement of the lower teeth during occlusion, the posttrite cones of the lower teeth could to some degree fit into the dentine basins of the pretrite cones of the corresponding upper molar (and *vice versa*). Thus, at least part of the complex masticatory function of the gomphothere dentition formed a mortar and pestle grinding structure. As the posttrite crest passed out of the pretrite basin it crossed the sharp enamel edge that bordered that basin, thus effecting some horizontal shearing. As wear progressed the grinding effect was lost and the worn enamel figures passed over each other, their sharp edges performing primarily the shearing function. The more worn anterior portion of the tooth row thus became at this stage an efficient shearing structure, while the posterior portion was still a combined grinding-shearing device. From this molar type arose the modifications that resulted in the elephantid dentition.

The dentition of the earliest elephants did not differ basically from that of the gomphotheres, but a number of important changes occurred which seem to reflect a functional shift in mastication. As the elephant tooth became longer, and was unaccompanied by any lengthening of the mandibular corpus, there was less room in the mandible. This resulted in fewer teeth in the jaw at any given time. The greater complexity of the milk molars functionally extended the tooth row forward and made efficient molars of the adult type available at an earlier age. Thus, the milk dentition increased in size and duration of use at the expense of the premolars. The latter were retained only in the earliest elephants.

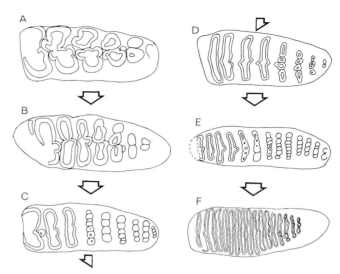

Fig. 23. Evolution of the elephant molar showing progressive consolidation of the gomphothere cone-pairs, loss of the median cleft, fusion of accessory columns, increase in plate number and thinning of enamel. A: *Gomphotherium angustidens*. B: *Stegotetrabelodon syrticus*. C: *Primelephas gomphotheroides*. D: *Mammuthus subplanifrons*. E: *M. africanavus*. F: *M. primigenius*. Not to scale.

As will be seen later, the predominant movement during the power-shearing stroke in elephants is a fore and aft one; scratches on worn molars confirm the lack of significant lateral movement, although some does occur. The major adaptive change that characterized the rise of the Elephantidae seems to have been a rapid reduction of the grinding component of mastication in favor of the shearing component (Maglio, 1972b). The lingual and buccal cones became anteroposteriorly compressed to form thin plates. The median cleft was obliterated by fusion of the cones, forming a single, continuous plate extending across the width of the tooth. The accessory columns responsible for the gomphothere trefoil wear pattern became partially incorporated into the plate face in elephants so that only a vertical ridge persisted. With wear, this ridge appeared as a loop on the enamel figure. In some later forms this ridge became completely incorporated into the plate (fig. 23). The transverse valleys between the gomphothere cone pairs broadened at the base in elephants, becoming U-shaped in all but the earliest elephants (fig. 24). The enamel became progressively thinner.

Functionally, this type of early elephant molar is understandable only in terms of the mandibular movements during mastication. Basically, these were of a fore-and-aft type with some lateral motion possible through differential movement of the condyles on the glenoid surface. As the jaw is brought forward in occlusion, the transverse enamel ridges on the partially worn surface of the lower molars pass over similar ridges on the upper molars. The leading edges of these enamel ridges form an effective shearing mechanism

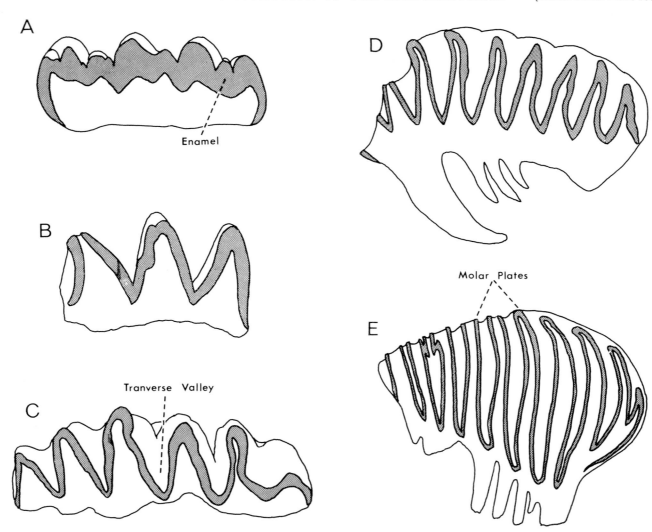

FIG. 24. Evolution of the transverse valleys between molar plates showing progressive narrowing and the change from V-shaped to U-shaped at the base. A: *Gomphotherium angustidens*. B: *Stegotetrabelodon orbus*. C: *Primelephas gomphotheroides*. D: *Elephas planifrons*. E: *Elephas maximus*. Not to scale.

analogous to the blades of a pair of scissors. Because the molars are curved, the uppers concave lingually and the lowers convex, the leading edges of the enamel ridges meet at a low, oblique angle that favors efficient shearing.[17] In general, as the jaw moves forward, the cutting angles between opposing shearing surfaces move outward on the anterior portion of the occlusal surface and inward on the posterior portion (fig. 25). This is true regardless of the lateral angular movement of the jaw during the shearing stroke. However, the curvature of elephant molars and the degree of obliqueness of the enamel ridges to the anteroposterior axis is variable enough so that the exact proportions of these shear directions is uncertain without detailed analysis of each specimen. It is unknown at present whether consistent differences occur in the various elephant groups.

This peculiar directional motion of shear points along the enamel ridges may serve in part to lengthen functionally the shear surfaces during any one forward stroke of the jaw. As food is cut on the anterior portion of the tooth, it is trapped within the buccally moving shear angles between crossed enamel ridges. At least some of this food would then come onto a more posterior portion of the tooth where the shear direction is lingual. Thus, the effective functional shear length may be longer than the actual length of the enamel ridges and may be attained by moving some of the food across the tooth, first buccally, then lingually, all in the same stroke. This structure seems also to provide a mechanism for directing food from the occlusal surface

[17] The leading edge for the lower teeth is the anterior surface of each enamel ridge; that of the upper molars is the posterior surface, as this is the one that first encounters the forwardly moving lower edges.

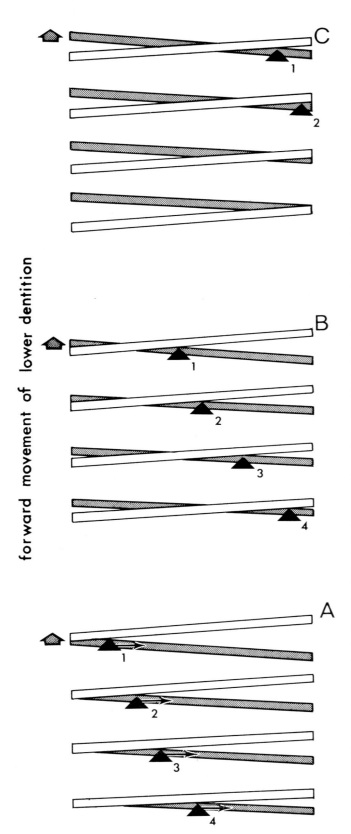

back into the mouth cavity for presentation to the teeth by the tongue before the next stroke.

In all lineages of elephants a single basic trend characterized the evolution of the dentition. This was the increase in total length of leading edges of enamel that form the shearing surfaces. All other modifications of the dentition seem to have been direct consequences of selection for this feature. There were three possible ways of accomplishing this increase: (1) broadening of the molar with attendant broadening of the enamel ridges; (2) folding of the enamel; and (3) increasing the number of ridges per unit distance of molar length. All three methods were utilized by the various groups, but especially the latter two methods. Broadening of the molar crown is a relatively simple solution to the problem, but the structural limitations of the skull and jaw seem to have prevented any major increase in width. Enamel folding was effective in several lines, but it was not sufficient in itself, and a combination of enamel folding and increased ridge number was used.

As in the case of crown width, the length of a molar apparently was severely limited by the architectural requirements of the skull and jaw. Thus, simple addition of plates on the posterior end of the molar by prolonging its growth period was not possible beyond a certain point. By reducing the spacing between adjacent plates, however, it became possible for most elephant groups to evolve molars with a greater number of plates without significantly increasing total length. This necessitated a basic alteration in the growth pattern of the tooth, since the closer spacing involved the anterior as well as the posterior plates, and not simply a lengthening of the growth period.

The shearing action of an elephant molar requires a certain separation of adjacent enamel ridges. If the ridges had become more closely spaced without other modification, the inter-ridge space would have been reduced, and overall efficiency lessened. Circumventing this, plates became thinner, the thinning affecting the enamel as well as the enclosed dentine. This process maintained the equal spacing of enamel ridges between alternating cement and dentine intervals. Since the shearing function depends solely on the leading edges of enamel, the reduced thickness did not impair masticatory efficiency. However, because the rate of molar wear depends upon the rate of erosion of enamel, the thinner enameled teeth of later elephants became subject to more rapid abrasion. This was compensated for by an increase in the crown height which in some

FIG. 25. Diagrammatic representation of shear angle movement during the forward thrust of the jaw. The cutting angles (1-4) between opposed upper and lower enamel ridges move across the molar as the teeth pass over each other. This results from the oblique angle at which the upper and lower enamel ridges meet. See text for full explanation.

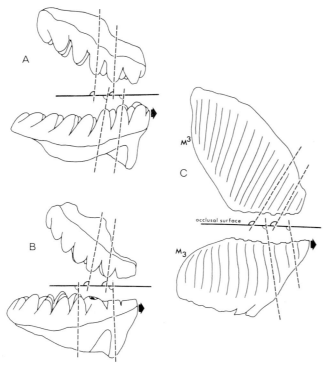

FIG. 26. Upper and lower molars of early and late species of elephants showing changes in the orientation of the plates with respect to the occlusal surface. The lower molar plates are directed forward in the earliest forms and backward in the latest species. This shift is presumably associated with reducing the rate of crown wear. A: *Stegotetrabelodon orbus*. B: *Primelephas gomphotheroides*. C: *Mammuthus primigenius*. Not to scale.

later species was six times as great as in the earliest forms.

The structure of the maxilla allowed great increase in the crown height of upper molars. But corresponding increase of lower molars seemingly was limited by the depth of the mandibular corpus which, for mechanical reasons, probably could not be greatly deepened. The method of growth of elephant teeth does not allow for an open root with continuously growing crowns as in some rodents. Each plate is fully formed at the posterior end of the molar crypt and added sequentially to the back end of the tooth. Thus, the entire height of the lower molars had to be accommodated within a mandible of limited dimensions. This restriction on the lower dentition but not the upper resulted in disproportionate increase between upper and lower heights, with the lower teeth proportionately lower-crowned in the more progressive species.

In apparent adjustment to this height disparity which would otherwise result in the wearing away of the lower molars before the uppers, the angle of occlusion of the enamel ridges was modified. In the gomphotheres and earliest elephants, the enamel ridges on the lower teeth were vertical in orientation or were inclined slightly forward. Those of the upper teeth were inclined slightly backward so that both upper and lower enamel surfaces faced each other like blades of two chisels (fig. 26). This tended to maintain sharp leading edges on the enamel ridges. It also led to rapid wear, which was compensated for in these early forms by the great thickness of the enamel.

In more progressive species in which the disproportionate crown height of upper and lower teeth was beginning to become manifest (e.g. *Loxodonta adaurora*), the enamel ridges of the lower molars are, in contrast, inclined backward at about 10 to 15 degrees to the occlusal plane. In still later species this backward inclination is even greater, 25 to 30 degrees in the living African elephant and 30 to 40 degrees in the Asiatic species. With this configuration the lower molar enamel ridges face *away* from the direction of jaw movement during shearing. In the absence of a crystallographic study of elephant enamel, the exact significance of this is uncertain, but it may have served to reduce the rate of erosion of these lower molar enamel surfaces so that both upper and lower teeth wore out at about the same time.

Other changes were also associated with the transition from the rounded trefoil-cones of the grinding-shearing gomphothere molars to the parallel-ridged, shearing ones of the elephants. Because of the partial mortar-and-pestle-like function of the gomphothere molar the upper and lower trefoil basins were on opposite sides of the occlusal surface so that both could function simultaneously. As the shearing surfaces were elaborated at the expense of the grinding basins in elelphants, the intravalley columns forming the trefoil folds became incorporated into the developing plates. Only small loops of enamel remained where the columns were incompletely fused to the plate faces. Concomitantly with the reduction of the trefoil columns and the development of plates, the median cleft was obliterated and symmetry was established about the remaining enamel loops. The process by which this symmetry was achieved involved the proliferation of minor crown divisions on the gomphothere cones to form the transverse row of partially fused pillars so typical of late gomphotheres and early elephants. An enamel "bud" from one of the centrally placed pillars (usually the one bearing the enamel loop) increased in height, expanded toward the midline and came to dominate its neighbors (fig. 27). In the process, it displaced the old median cleft, now merely a groove, to one side or the other. This new central pillar dominated the plate, forming the center of symmetry of the crown. The result was that the loops, representing the vestigial columns, came to occupy a more nearly central position on the worn enamel figures, and the upper and lower loops now came into direct occlusion without requiring significant lateral movement of the jaw. With even a small amount of lateral movement these loops provided additional shearing traps in the angles between them and the transverse enamel ridges. Such traps functioned by retaining food within a diminishing angular space bounded by shear surfaces. In those elephants in which the plates became very closely spaced, no room was left for the median loops and these disappeared completely. It is interesting to note that in these species the enamel is highly folded and apparently serves a similar function by providing a series of small shear traps.

An interesting, although simplified, measure of the shearing ability of an elephant molar may be calculated as follows:

$$\frac{2LF^3 \times 2LF_3 \times W_3}{1000},$$

where LF^3 is the lamellar frequency of the upper third molar, LF_3 is the lamellar frequency of the lower third molar, and W_3 is the maximum width of the lower molar, all measured in millimeters. This measures the total length, in meters, of the leading edges of enamel that shear past each other during a forward mandibular stroke of ten centimeters and may be called the shearing index. Since the lamellar frequency measures the number of plates per ten centimeter unit of molar length, twice that value gives the number of enamel ridges (two per plate). The width of the lower molar is used because it is narrower than the corresponding upper molar and is therefore the limiting factor in determining the occlusal width. The shearing index may be calculated for each species by using the mean values

TABLE 33

Shearing Index for M3 Calculated from the Mean Values of the Lamellar Frequency and Molar Width. The Index Is Defined as Follows: $\frac{2LF^3 \times 2LF_3 \times W_3}{1{,}000}$.

Species	LF^3	LF_3	W_3	Shearing Index
Stegotetrableodon syrticus	3.0	3.1	119.2	3.86
Stegotetrabelodon orbus	2.8	2.9	97.9	3.18
Primelephas gomphotheroides	3.1	3.3	94.5	3.87
Loxodonta adaurora	3.7	3.4	90.2	4.54
Loxodonta atlantica	4.2	4.6	89.1	6.89
Loxodonta africana	4.1	4.2	75.4	5.19
Elephas ekorensis	4.4	4.0	91.4	6.34
Elephas recki	5.1	5.0	90.6	9.24
Elephas iolensis	4.0	5.3	104.7	8.88
Elephas antiquus	5.6	5.7	75.1	9.59
Elephas planifrons	4.2	4.2	94.0	6.63
Elephas celebensis	7.9	6.2	42.5	8.33
Elephas hysudricus	5.1	5.4	91.7	10.10
Elephas maximus	6.6	7.2	77.0	14.90
Mammuthus subplanifrons	3.5	3.3	94.4	4.36
Mammuthus africanavus	4.1	3.8	91.9	5.37
Mammuthus meridionalis	4.6	4.9	97.2	8.76
Mammuthus armeniacus	6.3	6.5	87.6	14.35
Mammuthus primigenius	8.5	9.0	87.6	26.81

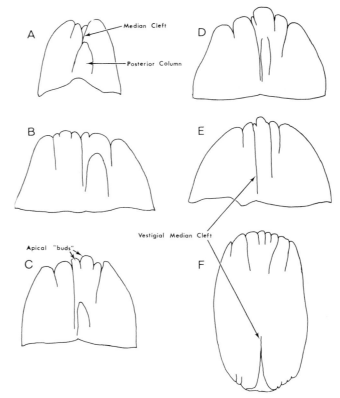

Fig. 27. Evolution of symmetry in the elephant molar showing development of "apical buds" and the displacement of the median cleft. The posterior column becomes incorporated into the plate face. A: *Gomphotherium angustidens*. B–D: *Stegotetrabelodon orbus*. E: *Primelephas gomphotheroides*. F: *Elephas recki*. Not to scale.

of the pertinent parameters for upper and lower molars. This gives a rough indication of the average index for the species (table 33).

The evolutionary trends seen in the various lineages of elephants with respect to these molar adaptations presumably reflect responses to the different ecological situations encountered by species involved. The last permanent molar is one of the most common elements in the fossil record (M1 and M2 often being shed before death) and it is important in the economy of the living animal because it serves for the major portion of the animal's life. The following comparisons will be restricted to this molar.

In all molar parameters *Loxodonta* was the most conservative group, both in terms of absolute change and of evolutionary rates. The number of plates on M3 did not change greatly between the earliest and latest species. An initial change is seen to accompany the emergence of the genus from *Primelephas gomphotheroides* to *Loxodonta adaurora* (fig. 28). From that point on, the change was not very impressive. As a result, such important characters as total enamel edge length and the correlated parameters of plate spacing,

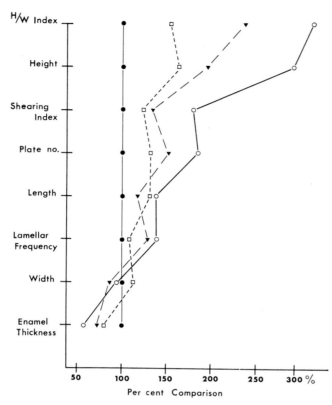

FIG. 28. Per cent comparison of molar characters for M^3 of three species of *Loxodonta* compared with *Primelephas gomphotheroides* as a standard. Solid circle: *P. gomphotheroides*. Open square: *L. adaurora*. Solid triangle: *L. africana*. Open circle: *L. atlantica*.

enamel thickness, etc., did not evolve to any striking degree through the nearly four million years that we can trace the genus. The most noticeable advance is in crown height, which in *L. africana* is about three times as great as in *P. gomphotheroides* and probably reflects some change in diet rather than strict compensation for a thinning of enamel. The absolute size of the teeth remained relatively constant in the genus except in *L. atlantica,* the giant of the group. The shearing index changed by about 25 per cent, from 4 to 5 meters, between *P. gomphotheroides* and *Loxodonta africana*. Compared to changes in other lineages, this is almost insignificant (fig. 29).

In the genus *Elephas* molar adaptations similar in kind are seen, but the number of plates more than tripled within the lineage leading to the living Asiatic elephant, *E. maximus*. As a result, the enamel thickness was reduced more than in *Loxodonta* and the median loop was reduced or lost in all lines. Intense enamel folding occurred during the later history of the genus, attaining its maximum in the living species. The compensatory changes in crown height resulted in a four-fold increase over that in *P. gomphotheroides* (fig. 30). The shearing index in *E. maximus* is nearly four times that of *P. gomphotheroides,* 15 meters *versus* 4 meters. This reflects a great increase in functional shearing length. In the lineages leading to *E. namadicus* and *E. iolensis,* increase in the number of plates was only about half that achieved in the *E. maximus* line. Yet, enamel was reduced nearly as much and the relative crown height increased to about the same extent. In both, the median loop was absent, but a somewhat pointed median expansion of the plates superficially resembled the *Loxodonta* pattern when worn. This probably represented a parallel development.

Dwarfing occurred several times independently within the *Elephas* complex. In the Malay Archipelago, Celebes received an overseas colonization of *E. planifrons* from southeastern Asia, probably during the early Pleistocene. *E. planifrons* itself is unknown on Celebes; presumably the immigrants underwent a very rapid reduction in size that resulted in the pygmy *E. celebensis*. Although this species was only one-half as large as *E. planifrons* in linear dimensions, the molar enamel thickness was reduced only by a few per cent (see tables 20 and 22). This evolution of proportionately thicker enamel is seen in dwarfs of other mammals (*e.g.* hippopotami, suids) and evidently serves to compensate for the rapid wear that would result from too great a reduction in enamel thickness without compensatory increase in absolute crown height.

In the Mediterranean area pygmy elephants were present on Sicily, Malta, Cyprus, and Crete in the Middle Pleistocene; all of them ultimately were derived from the European *Elephas namadicus*. In the smallest of them, *E. falconeri,* the molars have fewer plates and proportionately thicker enamel than in *E. namadicus*. This led to the speculation that it could not have been derived from that more progressive species. However, both the reduced plate number and the relatively thick enamel surely came about in direct response to the minute size of the molars (*E. falconeri* was only one meter in shoulder height). In comparison to body size, the molars are proportionately larger than in *E. namadicus*. The small skull and jaw could not easily accommodate molars with 13 or 14 efficiently spaced plates unless the enamel was reduced below critical thickness with respect to wear.[18] These modifications, then, probably are functionally related to the dwarfing process and should be used for assessing phyletic relationships only with great caution.

The molars of the *Mammuthus* lineage became highly specialized for a diet of tundra vegetation. As regards relative crown height, the lineage leading to *M. primigenius* closely paralleled that of *Elephas* (fig. 31). The number of plates on M3 surpassed that of *E. maximus* and the spacing of enamel ridges was proportionately less. In reduction of enamel thickness, *Mammuthus*

---

[18] It is true that certain rodents, such as capybaras, have up to 17 plates with thin enamel, but their molars are hypselodont, which no elephant molar ever was.

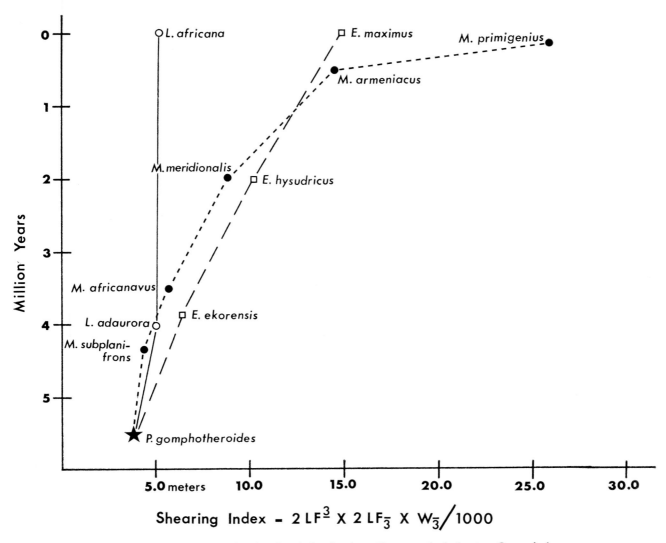

Fig. 29. Evolution of the shearing index in three lineages of elephants. Open circle: *Loxodonta.* Open square: *Elephas.* Solid circle: *Mammuthus.*

surpassed all other elephants. Because of close packing of the plates, the enamel ridges never became coarsely folded as in *Elephas,* although they were finely wavy. The molars were characteristically very broad. This, together with the closely spaced plates, gave *M. primigenius* a very high shearing index, one that was about seven times as great as in *P. gomphotheroides* (27 meters as compared with 4 meters; see fig. 29).

In summary, the molar evolution in all three lineages proceeded in closely parallel fashion for the first two million years of their histories. Thereafter, *Elephas* and *Mammuthus* were characterized by a very progressive development in all molar parameters. *Loxodonta* remained conservative throughout its history. In this connection, it is interesting to note that two of the more conservative lines, *Loxodonta* and *Elephas recki-E. iolensis,* never spread beyond the African continent, where climatic and ecologic conditions remained relatively more stable than elsewhere in the elephantine range. The most progressive lineages, *E. ekorensis-E. maximus* and especially *M. africanavus-M. primigenius,* expanded from Africa into northern continents where response to climatic deterioration and vegetational change during the glacial periods seems to have strongly influenced molar specialization (see pp. 111–118). This is most strikingly shown in the *Mammuthus* lineage, which was subjected to a greater fluctuation of climatic conditions than any other group of elephants.

EVOLUTION OF THE ELEPHANT MANDIBLE

It is not as easy to study evolutionary changes in the elephant mandible as it is in the dentition because of the relative scarcity of complete specimens, especially

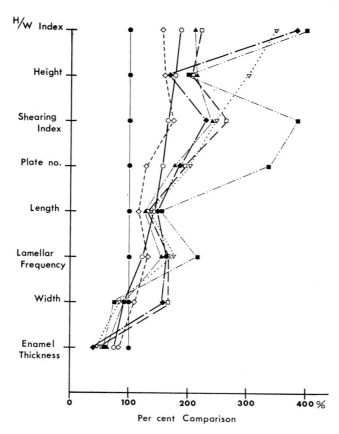

FIG. 30. Per cent comparison of molar characters for M³ of seven species of *Elephas* compared with *Primelephas gomphotheroides* as a standard. Solid circles: *P. gomphotheroides*. Open diamonds: *E. planifrons*. Open circles: *E. ekorensis*. Solid triangles: *E. recki*. Open squares: *E. hysudricus*. Open triangles: *E. namadicus*. Solid diamonds: *E. iolensis*. Solid squares: *E. maximus*.

In the earliest elephants the long symphysis and tusks were retained, but reduced in size. Specialization of the molars had resulted in a backward extension of the alveolar crypt so that the angle of the jaw was generally more massive than in *Gomphotherium*. In *Stegotetrabelodon syrticus* a strong platelike flange extended the angle backward. The ramus was broad anteroposteriorly, but did not extend anterior to the occlusal surface. The condyle was situated farther forward than in gomphotheres. The center of gravity was still far forward and the entire mass of the masticatory musculature lay behind the occlusal area of the dentition. In somewhat later forms, such as *Loxodonta adaurora* and *Mammuthus subplanifrons*, the mandibular incisors were lost externally, although superficially closed alveoli persisted. The symphysis was relatively short compared to earlier forms, but was longer and more protruding than in later species. The corpus was still rather long, but the ramus was higher and relatively farther forward than in *Stegotetrabelodon*. The coronoid lay over the posterior portion of the occlusal area and the condyle was more nearly above the cen-

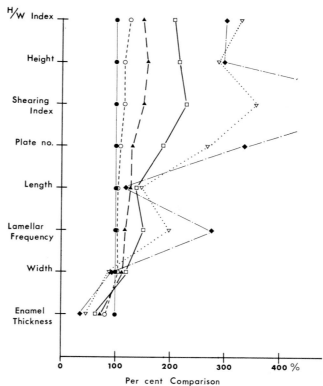

FIG. 31. Per cent comparison of molar characters for M³ of five species of *Mammuthus* compared with *Primelephas gomphotheroides* as a standard. Solid circles: *P. gomphotheroides*. Open circles: *Mammuthus subplanifrons*. Solid triangles: *M. africanavus*. Open squares: *M. meridionalis*. Open triangles: *M. armeniacus*. Solid diamonds: *M. primigenius*.

for the earlier species. From what is known, however, a number of striking changes in the overall structure of the jaw can be traced. These changes reflect certain fundamental specializations of the masticatory complex as a whole and can be correlated with modifications in the skull and in muscle configurations.

The mandible in *Gomphotherium* had a long symphysis with large incisors pointing forward and downward. The corpus was very long and relatively lightly built with a prominent angle. A flat, enlarged process on the angle provided attachment area for the superficial masseter externally and for the medial pterygoid muscle internally. This process was situated entirely behind the alveolus for M₃. The ramus was low and inclined backward, the coronoid lying almost completely behind the last molar. The condyle formed the posterior-most portion of the jaw. Because of the great weight of the symphysis and tusks, and the thin, lightly built ramus, the center of gravity of the jaw lay far forward near the anterior portion of the corpus.

tral part of the ramus. The angle was gently rounded and contained the posterior part of the molar alveolus. There was no flange.

With continued evolution of the masticatory apparatus during the Pleistocene, the symphysis and corpus became further shortened, and the ramus was extended anteriorly so that the coronoid process came to lie over the central portion of the occlusal surface. As the maxilla was depressed ventrally in response to the increasing height of the upper molars, the temporomandibular joint remained relatively fixed with respect to the skull, while the corpus and lower dentition were depressed with the upper jaw. The condyle thus became situated higher on the mandible. As a result of the elongation of the ramus, the forward extension of the coronoid process and the shortening of the corpus and symphysis, the center of gravity of the jaw was displaced backward and came to lie almost directly beneath the coronoid. In all lines the major evolutionary trend seems to have involved the transfer of the center of gravity backward to underlie the temporalis musculature. At the same time, changes in cranial architecture resulted in related changes in the mandible, affecting the condyle and muscle-bearing portions of the ramus.

The *Loxodonta* lineage was the most conservative group in terms of absolute change. The symphysis, even in the living species, is still prominent. The corpus has not been significantly reduced in length, but the coronoid is significantly more anterior in position. It is only slightly higher in *L. africana* than in *L. adaurora,* as would be expected since the upper molars did not greatly increase in height within this lineage (fig. 32). In *Elephas* we can only trace mandibular evolution for the last two million years, but here we see a greater modification than in *Loxodonta*. The symphysis was reduced more, the coronoid was shifted farther forward and the ramus was deepened resulting in a higher position of the condyle on the jaw. As was the case with the dentition, the greatest structural modifications are seen in *Mammuthus*. Here the corpus became very short and the symphysis tucked under and not protruding. The ramus and coronoid broadened from *M. subplanifrons* to *M. primigenius,* with the coronoid being very far forward in the latter. The condyle position became very high above the level of the dentition.

The function of these changes in jaw structure can only be understood in relation to correlated changes in the skull and muscle orientations and thus will be deferred until the section on the masticatory apparatus as a whole. It can be noted here, however, that although some changes seem to be correlated strictly with changes in the skull and dentition and with architectural limitations imposed by the mandible itself, the majority of changes in the jaw seem related to a reorganization

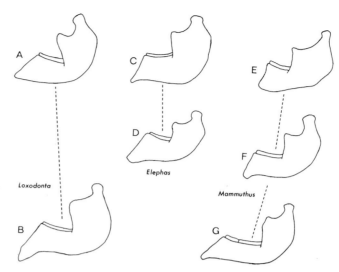

Fig. 32. Evolution of the elephant mandible in the three genera of Pleistocene elephants, showing progressive shortening of the corpus, broadening of the ramus and increased height of the condyle in later species. A: *Loxodonta africana;* B: *L. adaurora;* C: *Elephas maximus;* D: *E. hysudricus;* E: *Mammuthus primigenius;* F: *M. meridionalis;* G: *M. subplanifrons*. Not to scale.

of weight-force relationships between jaw and masticatory muscles.

EVOLUTION OF THE ELEPHANT SKULL

The structure of the elephant skull is unique among the Mammalia. Some of its specializations, such as the anteriorly placed orbits, the foreshortened basicranium, the posteriorly placed nasals and frontals, the enlarged tusks and premaxillaries, and the relatively forward position of the temporalis fossa were established early in the history of the Proboscidea. Others, such as the great elevation of the parietals, the more forward position of the occiput and squamosal bones, the extreme foreshortening of the skull, the downwardly directed premaxillaries, and the ventrally depressed palatal region were rapidly achieved after the emergence of the Elephantidae.

Weithofer (1890) regarded the specializations of the elephant skull as correlated primarily with the prodigious development of the tusks which required a reorganization of bone and muscle for their support. This idea was rejected by Gregory (1903), who argued that the development of the proboscis seemed to correlate best with the foreshortening of the skull in the elephant. Neither the tusks nor the trunk were innovations of the Elephantidae and, even though these structures did influence the evolution of the proboscidean skull in its earlier history, it seems unlikely that either played any major role in the marked modifications that accompanied the evolution of the Elephantidae. Large tusks and well-developed trunk were certainly present in the

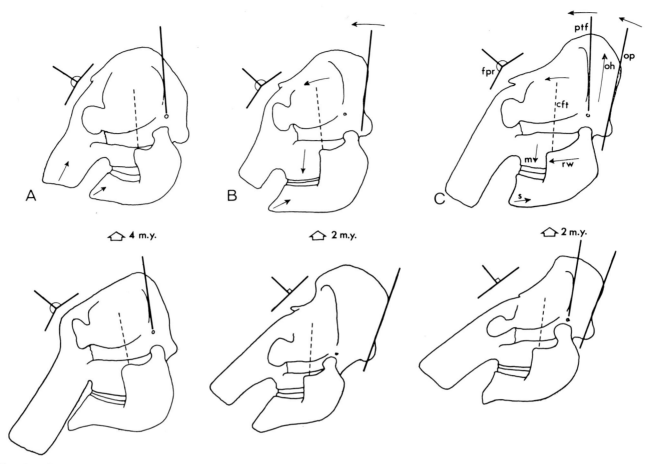

FIG. 33. Major changes in skull morphology occurring in the three genera of Pleistocene elephants. Abbreviations: *cft*, central fiber axis of the temporalis muscle. *fpr*, frontal-premaxillary angle. *m*, depth of the maxilla. *oh*, height of the occipital region. *op*, occipital plane. *ptf*, plane of the posterior limit of the temporalis fossa. *rw*, width of the mandibular ramus. *s*, mandibular symphysis. A: *Loxodonta*—below, *L. adaurora*; above, *L. africana*. B: *Elephas*—below, *E. hysudricus*; above, *E. maximus*. C: *Mammuthus*—below, *M. meridionalis*; above, *M. primigenius*.

mastodont, *Mammut*, but the skull is very different from that of the elephant. The differences reflect a basically different mode of mastication. Similarly, large tusks and a trunk were present in the stegodonts, but this group exhibits a great range of cranial morphologies. Some, such as in *Stegodon bombifrons*, closely resemble that of true elephants; others, such as *S. insignis* and *S. ganesa*, are widely divergent from all other Proboscidean skull types. Cranial adaptations in the stegodonts require study, but it is clear that a complex of factors control the morphology of the skull and these are not simply correlated with the presence or absence of tusks and a trunk.

In all later Proboscidea the forward position of the occipital plane, the development of the premaxillaries, the retraction and elevation of the frontal bones and the high position of the nasal aperture are common features that are clearly related to the presence of tusks and trunk. But the highly distinctive skull of the elephant was apparently the result of the specific masticatory adaptations that characterized the evolution of this group.

The skull of *Gomphotherium* shows a greater similarity to the elephant skull than does that of any other proboscidean, as we would expect from the phyletic relationship between these groups. By Miocene times the gomphothere skull already showed a marked elevation of the parietal region and a trend toward the more forward position of the temporalis muscle mass. The premaxillaries were oriented slightly more vertically than in mastodonts. The palate was narrow and the upper dentition of opposite sides was subparallel or slightly convergent anteriorly. The skull was long, with the orbit overlying the posterior part of the dentition. The entire masticatory muscle complex lay behind the occlusal surface of the dentition, including nearly all of the origin of the superficial masseter muscle. The temporalis muscle mass originated from a broad temporalis fossa that extended only slightly posterior to the glenoid surface.

TABLE 34

Per Cent Differences in Angular Displacement for "Central Fiber" Directions of Masticatory Muscles in *Loxodonta africana*. The Values Indicate the Per Cent Displacement in Angle of Pull between the Open and Closed Position of the Jaw at the Beginning of the Occlusal Phase (Jaw Retracted) and at the Beginning of the Recovery Phase (Jaw Forward) of the Masticatory Cycle. The Last Column for Each Position Compares the Displacement Resulting from the Utilization of a Pterygoid Pivot (See Text for Explanation) with that Resulting from a Condylar Pivot in Terms of the Per Cent Difference between Them. In Most Cases the Use of a Pterygoid Pivot Allows the Opening and Closing of the Mandible with Less Angular Displacement of Muscle Fibers while Maintaining the Same Degree of Movement of the Dentition.

Muscle	Jaw retracted			Jaw forward		
	Condyle pivot	Pterygoid pivot	Pteryg. vs. cond.	Condyle pivot	Pterygoid pivot	Pteryg. vs. cond.
Temporalis, anterior mass	5°	1°	80% L[1]	7°	2°	71% L
Temporalis, posterior mass	6°	0.5°	92% L	8°	2°	75% L
Superficial masseter	5°	2°	60% L	6°	2°	67% L
Deep masseter	16°	5°	69% L	16°	8°	50% L
Lateral pterygoid	1°	2°	50% G[2]	2°	2°	—
Medial pterygoid, dorsal mass	1°	3°	66% G	1°	7°	86% G
Medial pterygoid, ventral mass	1°	5°	80% G	4°	13°	69% G
Digastric	11°	5°	45% L	10°	5°	50% L

[1] "L" indicates that the pterygoid pivot displacement is *less* than the condylar pivot displacement by the given percentage.

[2] "G" indicates that the pterygoid pivot displacement is *greater* than the condylar pivot displacement by the given percentage.

seems to function primarily in providing a pivotal point around which the mandible may rotate in the sagittal plane during opening and closing. The condyle, as is required by a system of this sort, is completely mobile. As the jaw is opened, the corpus rotates downward and backward around the fixed point of the pterygoid insertion and the condyle rotates forward around the same point by sliding over the convex glenoid surface (fig. 36). The articular cartilage and capsular ligaments allow great mobility in this joint.

The functional significance of this type of joint is not yet completely certain. A pivotal point well below the condyle permits a smaller lateral displacement of the muscle-bearing ramus portion of the jaw while retaining essentially the same degree of mobility of the molar-bearing corpus. The result is that when the jaw is moved to any position in the masticatory cycle there is a minimum of angular displacement of muscle fiber directions, and a smaller degree of linear stretch in the muscle. This applies to most of the muscles of mastication (except the pterygoids), particularly the temporalis and superficial masseter. For these last two muscles, the degree of angular displacement of the central fiber direction is 60 to 90 per cent less with a pterygoid pivot than with a condylar pivot (table 34). Thus, the muscles can operate during every phase of mastication, with continuously changing mandibular positions, but minimum need to compensate for rapidly changing moment-arm relationships (fig. 37).

Because of the very short basicranial portion of the elephant skull, the digastric muscle originates on the post-jugular shelf just behind the glenoid surface. This muscle cannot act as efficiently in opening the jaw by applying torsional force about a pivot point as in other mammals because of its inefficient orientation. Its main function apparently is to elevate the condyle into the post-glenoid depression on the recovery stroke. This operation is crucial for placing the teeth in the proper position for firm occlusion during the next shearing stroke, and is related to the curved shape of the occlusal surface.

The masticatory cycle may be summarized in simplified form as follows (fig. 38):

Recovery phase—The mandible opens, by use of the digastrics; the superficial masseter, medial pterygoid, and temporalis muscles relax, while the lateral pterygoid remains taut, providing the pivot point around which the jaw rotates in the sagittal plane. As the jaw opens, the lateral pterygoid relaxes, slowly displacing the functional pivot point of the jaw backward while the condyle swings farther forward over the glenoid surface. Further contraction of the deep masseter and digastric muscles lifts the condyle into the post-glenoid depression.

Closing phase—With the condyles thus elevated, the lateral pterygoid contracts to a point where it once again provides a pivot for the ramus. Contraction of the temporalis muscle brings the teeth into occlusion while the condyle rotates backward over the glenoid surface and deep into the post-glenoid depression.

Shearing phase—With a constant occlusal force being provided by the temporalis muscle, the superficial masseter and the lateral and medial pterygoid muscles act tangentially to the occlusal plane, bringing the jaw for-

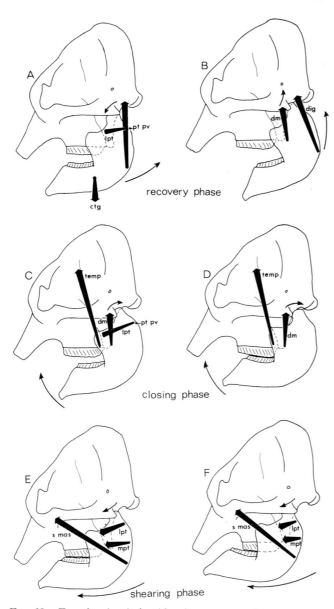

Fig. 38. Functional relationships between muscle masses during three phases of mastication in the elephant skull (*Elephas maximus* shown). Fine arrows indicate the direction of movement of the jaw and condyle; heavy arrows indicate the average direction of pull of muscle masses. *Recovery phase*: A, the lateral pterygoid is taut, providing the pivot point for the jaw; the jaw opens partially through gravity. B, the digastric aids in opening the jaw while it and the deep masseter elevate the condyle into the postglenoid depression. *Closing phase*: C-D, the lateral pterygoid provides a pivot for jaw rotation; the deep masseter and temporalis act radial to the occlusal surface, providing the occlusal force. *Shearing phase*: E-F, the medial and lateral pterygoids and the superficial masseter act tangential to the occlusal surface, providing the shearing force. Abbreviations: *ctg*, center of gravity. *dig*, digastric muscle. *dm*, deep masseter muscle. *lpt*, lateral pterygoid muscle. *mpt*, medial pterygoid muscle. *pt, pv*, pterygoid pivot point. *s mas*, superficial masseter muscle. *temp*, temporalis muscle.

ward along the arc of the occlusal surface. The condyle describes a smaller arc as it moves passively with the dentition. Throughout the shearing phase the temporalis muscle remains essentially radial in orientation to the occlusal plane. Some lateral movement is achieved by the differential forward and backward movement of the rami by contraction of the deep masseters and the pterygoids.

We may now turn from this brief analysis of the functional relationships of the elephant jaw during mastication to a further consideration of the fossil evidence. When viewed as a functional unit, the various changes that characterized the evolution of the skull and dentition in the different elephant lineages are more easily understood. All lines were evolving in roughly parallel fashion as they perfected the elephant type of feeding adaptation to various degrees. The details of specialization, as we have seen, differed in each line. These differences were merely modifications superimposed on the broader needs of the elephant adaptive zone.

The basic cranial modifications seen in all lineages may be interrelated as follows:

1) As the shearing ability of the elephant molar was increased, the crown height became greater in order to compensate for more rapid wear. This required a deeper palate for accommodation of the higher molars. As the maxilla was lowered, so the mandible was depressed, but the stable architecture of the skull base seems to have been responsible for the dentary-squamosal joint remaining relatively fixed in position. In order to compensate for this elevated position of the condyle and the attendant consequences on rotational displacement of muscle fibers about the joint, the ramus increased in height and the lateral pterygoid shifted its insertion to a position lower on it. This new insertion functionally replaced the condylar joint as a point of rotation for the jaw. The ramus was then free to continue its increase in height without undue rotational displacement of important masticatory muscles.

2) The forward displacement of the occipital and parietal regions shifted the temporalis musculature anteriorly to a position directly above the dentition and along a radius of the arc formed by the occlusal surface. This radial position allowed direct force to be exerted on the entire occlusal surface without creating simultaneous backward stress on the condyles.

3) With continued shortening of the symphysis, the center of gravity of the jaw was shifted backward. This was accompanied by a simultaneous forward expansion of the ramus to a point above the center and beneath the central fiber axis of the temporalis muscles. The result was a perfectly balanced mandible without torsional stress caused by gravity. There being no need to overcome this type of torsion, the superficial masseter was enabled to function more in the forward shearing stroke. The upward and forward orientation

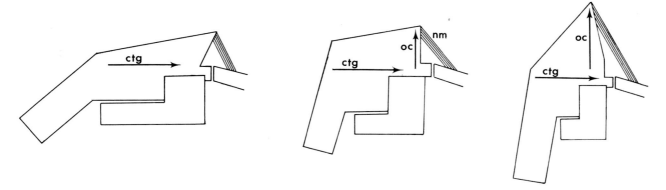

Fig. 39. Generalized, diagrammatic representation of changes in the elephant skull associated with modifications in weight relationships and the position of the nuchal musculature. As the center of gravity of the skull is shifted backward, the insertion of the nuchal muscles is elevated above the level of the occipital condyles. Abbreviations: *ctg*, center of gravity. *nm*, nuchal muscles. *oc*; occipital region of the skull.

of the superficial masseter can alleviate stress at the temporomandibular joint when occlusal force is restricted to the anterior portion of the dentition, in front of the temporalis insertion.

4) As the occipito-parietal region moved forward in relation to the occipital condyles, the moment-arm efficiency of the nuchal musculature was reduced. Possibly in compensation for this, at least in part, the occiput was elevated and the skull as a whole compressed anteroposteriorly through downturning of the tusks and face. This resulted in a posterior shift of the center of gravity of the entire skull, thus giving a more favorable mechanical advantage to the nuchal muscles (fig. 39). Although evidence is lacking, differences in use of the tusks may have strongly influenced their relationship to the skull as a whole.

VIII. RATES OF EVOLUTION IN THE ELEPHANTIDAE

The absolute rate of morphological change differed between phyletic lines and varied in tempo between one segment and another within each line. As has already been stated above, the *Loxodonta* lineage evolved very slowly in all characters over the four million years of its known history (fig. 40 to 42). The shearing index of the dentition increased as much in the million years from *P. gomphotheroides* to *L. adaurora* as it did in the whole of the subsequent four million years of evolution from *L. adaurora* to *L. africana*. The shearing ability increased somewhat faster in the line leading to the middle Pleistocene *L. atlantica*, but this was still a conservative change in comparison with that of other elephant groups. In other molar features such as enamel thickness and crown height, a similar retardation in rate of change is seen in the later history of *Loxodonta*. Since most of the cranial specializations of this group were established early, these also show remarkably slow modification during the late Pliocene and Pleistocene. It would seem that, once established, *Loxodonta* underwent little directional selection except for minor refinements in its dentition and skull.

In contrast to *Loxodonta*, the *Mammuthus* lineage underwent a remarkable burst in evolutionary activity

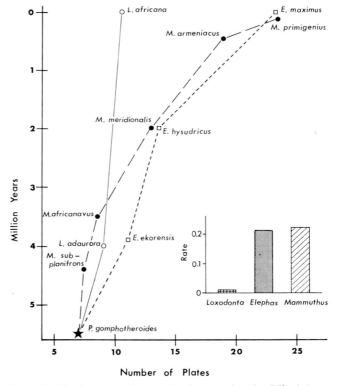

Fig. 40. Evolutionary changes in plate number for M^3 of three lineages of elephants. The absolute rate of change, measured in *darwins*, is given for the entire lineage as represented. Abbreviations: open circles, *Loxodonta*. Open squares, *Elephas*. Solid circles, *Mammuthus*.

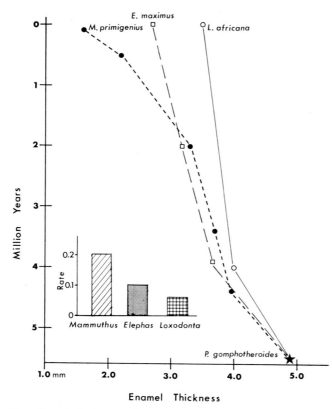

Fig. 41. Evolutionary changes in enamel thickness for M³ of three lineages of elephants. The absolute rate of change, measured in *darwins*, is given for each lineage as represented. Abbreviations: open circles, *Loxodonta*. Open squares, *Elephas*. Solid circles, *Mammuthus*.

tional adaptations of the dentition and jaw. The lower tusks were lost and the molars were already well evolved in *E. ekorensis*. Once this structural shift had been achieved, relatively slow changes characterized the lineage through the entire early and middle Pleistocene. In the later Pleistocene another rapid phase of molar evolution resulted in the emergence of the more progressive *E. iolensis* in North and South Africa (fig. 44). Although it is too early to speculate on the precise causes of this late burst in rate, it may be noted that the emergence of *E. iolensis* roughly coincided with the expansion of *Loxodonta atlantica* in the same regions. It is uncertain whether or not the latter was in direct competition with *Elephas*, but at several North African localities both species are recorded together.

In Eurasia, *Elephas namadicus* appeared suddenly in the middle Pleistocene, apparently derived from the African *E. recki* group sometime during the early Pleistocene. The exact time and place of origin for this species is still uncertain, but it must have arisen very rapidly in its initial phases, followed by a relative retardation during later stages. After its emergence the species changed very little except for the derived dwarf populations in the Mediterranean. The stratigraphic data are insufficient to permit the determination

in the later stages of its history (fig. 43). For the first three million years of its known record the teeth gradually increased their shearing capability, changing several times faster than in *Loxodonta*. The mandible shortened slowly; in *M. meridionalis* the corpus is shorter and the symphysis more retracted than in *M. subplanifrons*. During the middle Pleistocene all molar characters underwent rapid change (fig. 43), as did the shortening of the mandible, the forward rotation of the occiput and the downward flexure of the tusks. This marked increase in rate coincided with the appearance in Europe during the Cromerian of a probable competitor, *Elephas namadicus*. It also coincided with a major advance of glacial conditions in northern Europe. The intense selection pressures resulting from vegetation changes associated with this climatic deterioration and the apparent replacement of *Mammuthus* by *E. namadicus* in the warmer, more forested areas of southern Europe apparently led to a rapid adaptation to sub-arctic conditions in the mammoth line.

In the African branch of the *Elephas* complex a rapid rate of change in molar structure is seen in the initial stages from *P. gomphotheroides* to *E. ekorensis*, a transition representing an important shift in the func-

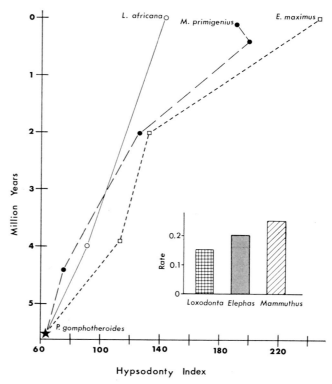

Fig. 42. Evolutionary changes in hypsodonty index for M³ of three lineages of elephants. The absolute rate of change, measured in *darwins*, is given for each lineage as represented. Abbreviations: open circles, *Loxodonta*. Open squares, *Elephas*. Solid circles, *Mammuthus*.

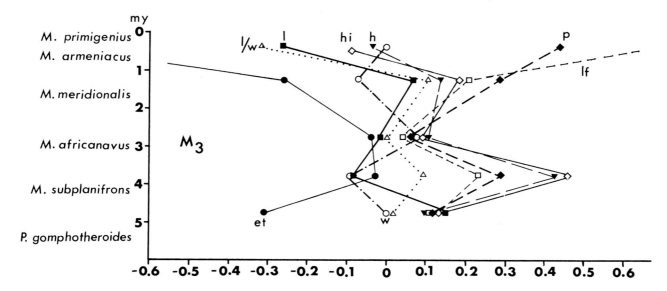

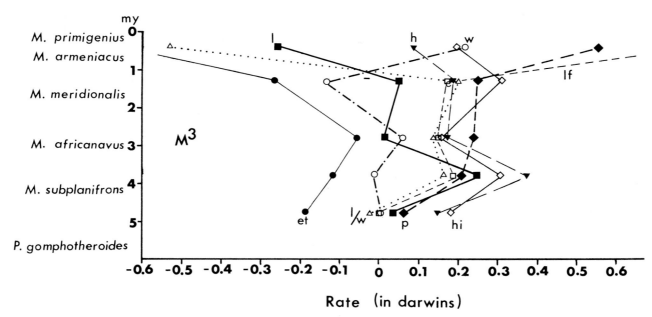

Fig. 43. Rates of evolutionary change, in *darwins*, for eight molar characters calculated for each successive species interval within the lineage *Primelephas gomphotheroides—Mammuthus primigenius*. One darwin is the change equal to a factor of *e* per million years. Negative rates indicate a decrease in the value of the measured parameter. Abbreviations: *et*, enamel thickness. *h*, height. *hi*, hypsodonty index. *l*, length. *lf*, lamellar frequency. *l/w*, length/width index. *p*, number of plates. *w*, width.

of accurate rates here, but the dwarfing process—some 4/5 reduction in linear dimensions for *E. falconeri*—appears to have taken place in a very short period of time, perhaps several hundred thousand years at most.

The Asiatic branch of the *Elephas* group did not experience the evolutionary stagnation seen at times in other lines within the genus. Here, a relatively steady, high rate of change in molar and cranial parameters occurred from *P. gomphotheroides* to *E. maximus*. The rate of increase in the molar shearing index was about five times that in the *Loxodonta* lineage. Expansion of the occipital region, downturning of the tusks and forward displacement of the temporalis musculature seems to have progressed more or less steadily. A more rapid rate of increase in molar plate number and height is seen in the transition from *E. hysudricus* to *E. maximus*.

If we now look at the pattern of evolution in the

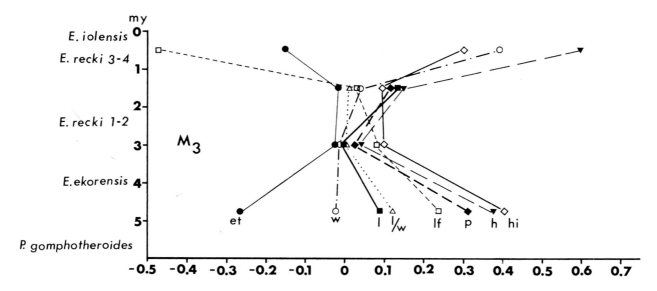

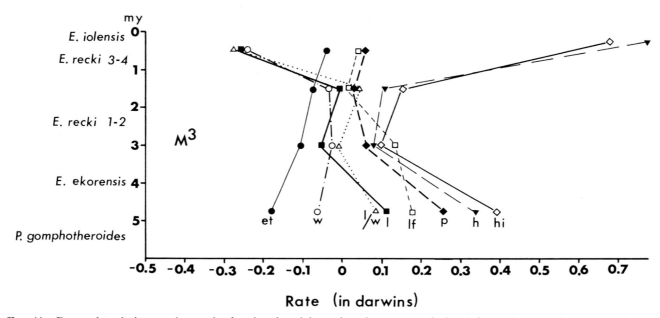

Fig. 44. Rates of evolutionary change, in *darwins*, for eight molar characters, calculated for each successive species interval within the lineage *Primelephas gomphotheroides-Elephas iolensis*. Abbreviations as in fig. 43.

family Elephantidae as a whole, a number of interesting phenomena emerge. As was seen in figure 15, the family was still diversifying until nearly the close of the Pleistocene, and there was little extinction during this time. It was suggested that this pattern reflects a group undergoing active adaptive radiation and that the more or less simultaneous extinction of all but two lines in latest Pleistocene times was the result of external causes. Such a pattern is even more apparent when an analysis of species diversification in each succeeding time unit is made.

In table 35 the six-million-year known history of the elephant family is divided into six one-million-year units. Numbers of first and last appearances and of persistent carry-over species are given for each unit as are the number of species that became extinct in each. Two additional columns show the numbers of the first appearances that arose by phyletic evolution and that arose by speciation (phylogenetic branching). The "turnover rate" represents the percentage of the total number of species in each time unit that appear for the first time in that unit; a value of 100 per cent indicates

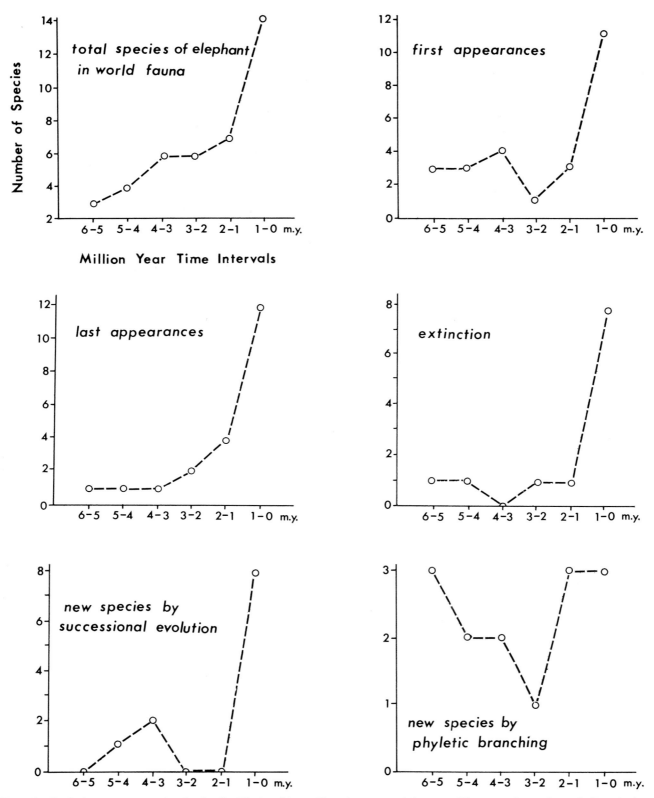

Fig. 45. Evolutionary patterns in the family Elephantidae. Time is measured in one-million-year intervals as indicated along the abscissa. Each point is calculated as the total number of species or events for the entire time interval. See table 35 for complete data.

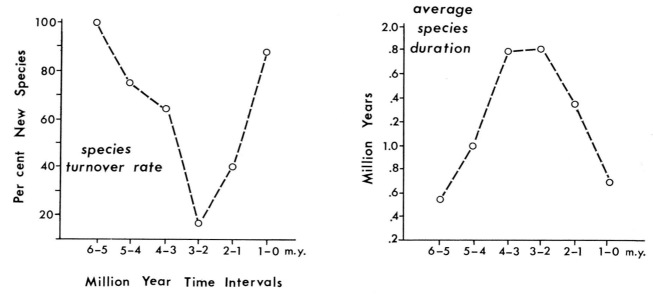

Fig. 46. Turnover rate and average species duration for each of six one-million-year time intervals in the evolution of the Elephantidae. See table 35 for complete data.

complete turnover with no survivors from the previous time period. The final column gives the average species duration in millions of years calculated as the average duration of all species occurring in that time unit. Because the values are given for one-million year intervals, the first and last appearances, turnover, extinction, and speciation data also represent absolute rates in numbers of species per million years. Figures 45 and 46 show the same data graphically.

TABLE 35

EVOLUTIONARY PATTERNS IN THE ELEPHANTIDAE SHOWING NUMBERS OF SPECIES MAKING THEIR FIRST OR LAST APPEARANCES, BECOMING EXTINCT, ETC. IN EACH OF SIX ONE-MILLION-YEAR INTERVALS. SEE TEXT FOR EXPLANATION.

	Time intervals (million years)					
	6–5	5–4	4–3	3–2	2–1	1–0
Total no. of species	3	4	6	6	7	14
First appearances	3	3	4	1	3	11
Last appearances	1	1	1	2	4	12
Carry-through species	0	1	2	4	4	3
Lineage extinctions	1	1	0	1	1	8
Arising by succession	0	1	2	0	0	8
Arising by speciation	3*	2	2	1	3	3
Turnover rate	100%	75%	66%	17%	43%	86%
Average species duration (m.y.)	0.5	1.0	1.8	1.8	1.4	0.7

* These probably arose somewhat prior to this interval but their early histories are unknown. It is clear, however, that by this time they represent very closely related independent lineages.

The most apparent observation is that the number of species recognized within the family increased continuously, especially during the Pleistocene when the total number doubled during the last one-million-year period. Although this certainly does represent a true diversification, it is important to determine the basis of this apparent increase in species number. If we calculate the number of first appearances in each time unit, regardless of whether these resulted from speciation or phyletic evolution, we find a similar increase during the Pleistocene. This increase was immediately preceded by a relatively low rate of new appearances during the time period 3.0-2.0 m.y. This retardation in first appearances is not seen in the total species number curve because during this period a larger number of persistent species carried through from earlier times. The overall pattern of the first appearances curve indicates a relatively steady rate of about three to four new species per million years for the first half of the family's history. This was followed by a retardation in late Pliocene times and a subsequent increase to eleven species per million years during the later Pleistocene.

The number of species making their last appearance in each time unit remained relatively low except for the last two million years. Some of this apparent "extinction" actually is due to the successional transformation of one "species" into another. This is clearly seen in the curve for rate of successional species appearance in which the last one-million-year period is characterized by a large number of species that were derived as successional stages of pre-existing lineages.

If only terminations of lineages are counted, an

extinction rate curve is obtained as in figure 45. Through most of the history of the family this rate was zero to one species per million years, but during the later part of the Pleistocene the rate soared to eight species per million years. It is particularly interesting to note here that within the subfamily Elephantinae only two true lineage terminations occurred prior to the mass extinctions of the late Pleistocene.

The number of true speciations or phyletic branchings may be estimated from figure 15 by counting the number of new branch lineages arising in each time unit during which the parental stock persists. The resulting rate distribution (fig. 45) indicates a high speciation rate, particularly during the early history of the family and again during the late Pleistocene. As with successional species, a relative retardation of branching is observed between 3.0 and 2.0 m.y. This retardation in new species appearances had a drastic effect on the turnover rate. The latter (fig. 46) shows a high rate of species replacement during the early evolution of the family and again during the Pleistocene, but a very slow turnover rate is seen between 3.0 and 2.0 m.y. This was true for all three genera as well as for the family as a whole.

One final measure of evolutionary activity is that of species duration. This calculation must be viewed with great caution, however, because of the arbitrary nature of lineage segmentation. Nevertheless, the results shown in figure 46 are compatible with the rates of diversification already discussed. It can be seen that the early history of the family as currently known was characterized by relatively short-lived species in which the average duration was about 0.5 m.y. A similar period of short-lived species occurred during the later Pleistocene. In the late Pliocene and early Pleistocene, species durations averaged 1.5 to 2.0 million years. This reflects a slower rate of evolution within each lineage as was discussed earlier (figs. 43 and 44).

Considering all these data together an interesting pattern of evolution emerges for the family. Shortly after the origin of the Elephantidae only a few species were present at any one time, extinction rates were low and the turnover rate was high as new species appeared primarily through speciation. With the emergence of the more progressive Elephantinae the rate of turnover remained high and the number of new species increased as persistent phyletic evolution became an important factor in diversification through time. The average duration of a species increased, but remained relatively short. Extinction rate was still very low. During the late Pliocene and early Pleistocene major expansion of geographic ranges was taking place. Numbers of species continued to increase with no extinction. The turnover rate was very low as species underwent stabilizing selection. Species duration was very high and morphological change was slow. The number of new species appearing was very low and all of these resulted from speciation, primarily in Eurasia.

A second rapid phase of evolution and radiation began during the middle Pleistocene. A spectacular increase in species number was accompanied by a rapid turnover rate resulting from a high rate both of speciation and of phyletic evolution. Average species duration fell rapidly so that by the late Pleistocene short-lived species were again the rule. The rate of morphological change along each lineage was very great. The reasons for this late radiation in Asia are still uncertain, but in Europe competition between *Elephas* and *Mammuthus* and ecological changes accompanying the onset of intense glacial conditions certainly must have had their effects. The breaking up of the environment into islands and isolated refugia also offered new opportunities for speciation.

Suddenly, in the midst of this radiation unparalleled in the Proboscidea, the extinction rate rose to a phenomenal degree, to eight species per million years during the late Pleistocene. The result was the termination of all but two lines, one in Africa and one in Asia. Even the continued existence of these is in doubt. The evolutionary pattern exhibited by this family is not one that suggests phyletic "senescence." Rather, it is one of active diversification and there is no reason to believe that the family had attained its maximum potential. The more or less synchronous extinctions in the late Pleistocene strongly suggest some external causation. In this regard the activities of man, with his ability to alter delicate ecological balance, cannot be discounted, although certainly other factors must also have been involved. (For an extended discussion of this problem see Martin and Wright, 1967 and Van Valen, 1971.)

IX. ZOOGEOGRAPHY

With the phyletic relationships of the family outlined, and with a chronology of Pliocene and Pleistocene deposits in which fossil elephants occur set forth, it is desirable to outline the zoogeographic events that led to the nearly world-wide distribution of the Elephantidae during the Pleistocene.

Before proceeding to the wanderings of Pleistocene elephants, it is advisable to sketch briefly, and only in the very broadest terms, the geological events that took place in the Mediterranean basin and surrounding areas during the Pliocene and Pleistocene. This region was critical for communication between Africa and Eurasia, and therefore a consideration of possible routes open to mammals, as well as the relative availability of these routes during this time, will be helpful in later discussions.

The early Miocene Tethys seaway joined the Atlantic through southern Spain and Morocco, producing a broad water barrier across what is now the Strait of Gibraltar. The absence of typically African species in the early Miocene of Spain (Savage, 1967) reflects the efficiency of this barrier. Transgressive seas of this period completely cut off Europe from North Africa except for a route across the Middle East. Although the Red Sea Graben had already become flooded as far south as the Farasan Islands (Swartz and Arden, 1960), faulting in the Aden Gulf had only just begun, and a corridor nearly 400 miles wide still connected what is now Eritrea to the Yemen Coast on the Arabian peninsula; from there the way was open to Asia and Europe.

By middle Miocene times, crustal movements and folding in parts of the Betic-Riff Cordillera of southern Spain and North Africa was still in progress (Fernex et al., 1965); somewhat later the Mediterranean was cut off from the Atlantic Ocean, and a land connection established between Spain and Morocco (Mattauer, 1963).

Renewed transgressions during the late Miocene were only partially offset by continued orogenic movements of the Apennines and Alps, and by uplift of the Middle East region. Maximum transgression was 100 to 500 meters above present sea level (Gignoux, 1955) making trans-Mediterranean sweepstakes routes via Sicily less likely than at other times. By latest Miocene time the Isthmus of Suez was fully emergent. The Aden rift was widening, but still not joined to the Red Sea Graben; thus a second corridor was still available in this area. The region of present-day Greece was emergent with continental deposition in the large Aegean, Dacian, and Panonian basins. A broad terrestrial corridor was thus open from northern Africa to Europe via the Middle East and the Aegean, and to Asia via Arabia and the coastal belt south of the Iran Plateau.

At the western end of the Mediterranean basin, early Pliocene faulting in the central parts of the Betic-Riff Massif resulted in the opening of the present Straits of Gibraltar. Recent coring operations of the Glomar Challenger have revealed salt accumulations laid down in a highly saline Mediterranean Sea, suggesting persistence of a continuous Betic-Riff connection across the Straits until the early Pliocene, perhaps five million years ago (Hammon, 1970).

Climatic conditions in the early Pliocene were cool, with pine, spruce and fir forests in the Balkans, Italy, and the Caucasus. Subarctic parklands and forest-steppe conditions extended across most of northern Eurasia, with steppe and desert habitats east of the Caspian Sea and north of the Himalayas. India probably maintained a mixed southern forest of pine, spruce and beech. Glacial conditions probably existed in the Alps and Scandinavia (Frenzel, 1968). In East Africa, open savannah of the present type, probably with greater rainfall, and broad gallery forests apparently dominated the eastern rift areas of Kenya and Ethiopia. Moist climates with forest cover, alternating with dry periods, characterized the late Tertiary of the Chad basin (Franz, 1968). North Africa was probably similar to southern Europe as regards its vegetation and climate.

Continued uplift and faulting in the Middle East during the early Pleistocene flooded the downfaulted Aqaba Graben, which became connected to a series of interior basins along the Jordan Valley. Further widening of the Aden Gulf cut off the southern corridor to the Arabian peninsula by opening the Bab el Mandeb Strait. Progressive falling of Pleistocene sea levels left the narrow Isthmus of Suez as the only connecting link between Africa and the Middle East (Swartz and Arden, 1960).

Subsidence of the Aegean Sea region and opening of the narrow Dardanelles in the early Pleistocene cut off the formerly broad corridor to Europe (Kummel, 1970), but this probably did not have a significant effect on mammalian movements.

Middle and late Pleistocene alterations in sea level accompanied the advances and retreats of ice sheets in the north. A preglacial sea level of approximately 100 meters above present level is often assumed, with subsequent interglacial levels progressively lower (e.g. Zeuner, 1959). However, this apparent lowering of sea levels may, in part, be due to epeirogenic uplift of continental masses, as suggested for the Gibraltar area by Giermann (1962). It is, therefore, not possible to reconstruct possible land routes during the later Pleistocene except for the last glacial and interglacial when sea levels can be measured accurately. During the last glaciation in Europe, sea levels were lowered by about 100 meters, widening the Suez Isthmus. The gap at Gibraltar was widened to approximately six miles and that between Sicily and Tunisia to a minimum of 11 miles. Thus, a sweepstakes route certainly existed in the western Mediterranean, at least intermittently, during the late Pleistocene and possibly during the middle Pleistocene as well. High evaporation from the great seas of the Caspian region during interglacial phases resulted in regression of shorelines, providing broader areas of access to eastern Europe. Uplift of mountains (except the Alpine belt) of some 200 to 300 meters resulted in an increased difference between the periphery of the continent and its center, with tremendous local variability of climates (Frenzel, 1968), which no doubt made possible the survival of some species in refugia.

In summary, the major possible expansion routes were as follows:

Pliocene: Between Africa and Asia via Suez and Aden; Africa and Europe via Suez, the Aegean region, and Gibraltar.

Plio/Pleistocene: Between Africa, Europe, and Asia by a narrow route at the Suez Isthmus.

Pleistocene: Between Africa and Asia via Suez; Africa and Europe via Suez, Gibraltar, and Sicily.

The earliest true elephants, the Stegotetrabelodontinae, seem to have been confined to the African conti-

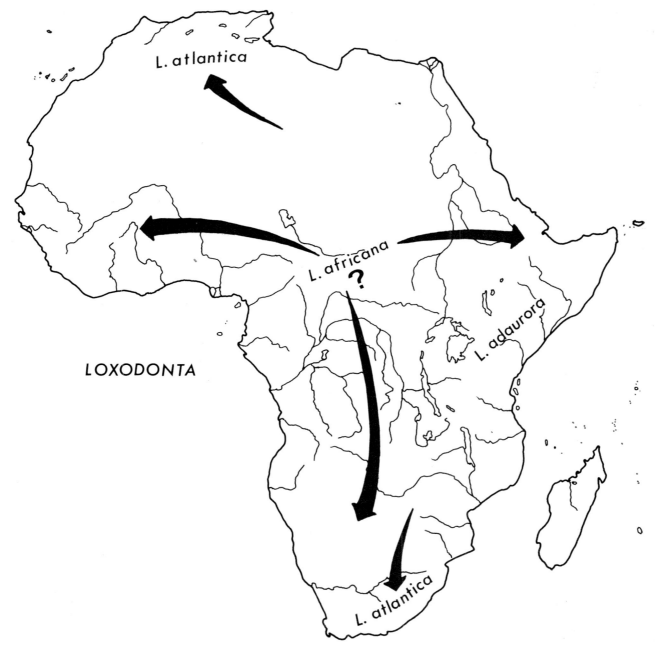

Fig. 47. Proposed geographic expansion for species of the genus *Loxodonta* in Africa. Arrows are not intended to represent exact paths. Places of origin for individual species remain uncertain.

nent where the group arose from the Gomphotheriidae probably during late Miocene times. The absence of this group from southern Africa is probably only apparent, owing more to the absence of suitable deposits of this age than to any geographical barrier to dispersal.

In the latest Miocene of eastern and northern Africa at least three species of *Stegotetrabelodon* are represented. From this and other faunal evidence (suids, brachypotherine rhinoceroses), the Sahara does not appear to have been a persistent barrier to mammalian expansion during the early Pliocene. It would appear that soon after its origin this elephantid group quickly rose to dominance in Africa, apparently replacing earlier gomphotheres with the exception of the Anancinae. Although expansion routes seem to have been open, this group did not gain access to Europe or Asia.

The earliest known stage of the Elephantinae (*Primelephas*) is represented only in East and Central

Africa. This may be due to the lack of deposits in other parts of the continent, or it may reflect a brief existence of a more or less localized transitional group from the early stegotetrabelodonts to more progressive elephants. The group has not been recorded outside of Africa, nor is it known north of the Sahara, but there deposits of appropriate age are unknown.

The earliest member of the *Loxodonta* lineage is East African in distribution. Some specimens of uncertain reference are known from South Africa and may be closely related. The genus appears to have spread quickly throughout sub-Saharan Africa during the late Pliocene and earliest Pleistocene. It is unknown from latest Pliocene beds at Ichkeul, Kebili, and Aïn Boucherit, and does not appear to have reached North Africa until the middle Pleistocene where it is well represented at Ternifine and later deposits in Morocco.

Loxodonta adaurora is the earliest species of the genus in East Africa, persisting into the late Pliocene of Ethiopia and Kenya, but becoming rare in comparison with its former abundance, even in areas of apparently unchanged ecology. It seems likely that it was replaced by *Elephas recki* in East Africa.

By early Pleistocene time, a primitive stage of the more specialized *L. atlantica* had arisen; it is known only from the Shungura Formation in southern Ethiopia. Soon after its emergence, this species apparently expanded into northern and southern Africa, disappearing completely from the equatorial regions. Its disappearance, and the concomitantly reduced occurrence of *L. adaurora,* coincides with the rise of *E. recki* in the same areas. *L. atlantica* again appears in the middle Pleistocene of Ternifine and Elandsfontein as a slightly more progressive stage, becoming the dominant elephant in both North and South Africa at this time. With the renewed late Pleistocene expansion of *Elephas*, in the form of *E. iolensis,* however, *L. atlantica* disappeared entirely. Only with the extinction of *Elephas* in Africa during latest Pleistocene time, did *Loxodonta* re-emerge as the dominant species on the continent. This genus seems never to have expanded beyond the continental limits (fig. 47).

The failure of *Loxodonta* at intercontinental expansion seems to be due primarily to the fact that its distribution during most of the earlier Pleistocene excluded it from North Africa. In middle and late Pleistocene time, probably only a sweepstakes route was open to Europe, via Sicily. Deteriorating climatic conditions in Europe at this time may also have proved a more serious barrier against northward expansion for this particular group.

Like *Loxodonta*, the earliest recognizable species of the genus *Elephas*, *E. ekorensis*, is known from the early Pliocene of East Africa. It seems almost certain that the genus did not enter northern Africa until late in the Pleistocene epoch. The time of arrival in southern Africa would appear to have been earlier; a form close to *E. ekorensis* occurs in the Chiwondo Beds, at Sterkfontein and Bolt's Farm in the Transvaal.

Elephas ekorensis appears to have evolved directly into *E. recki* in East Africa, with the latter spreading northward, but not southward, during the late Pliocene. The *recki* grade is represented in Chad, but is unknown in North Africa. The latest stage in this specific lineage, *E. iolensis*, spread rapidly over the entire continent, occurring from Cape Town to Algeria during the middle and late Pleistocene.

On present evidence, Africa appears to have been the center of origin and dispersal of the Eurasian branches of *Elephas* as well as of the *E. recki* line. In the late Pliocene, a form close to *E. ekorensis* expanded northward across either the Aden or Suez corridors into the Middle East region, where it is found in the Bethlehem fauna of Jordan (fig. 48). Piccard (1937) rejected the earlier held theory of faunal communication between Africa and Asia during the Plio/Pleistocene because of arid climatic conditions in the Palestine-Sinai and East Jordan area. Certainly, progressively more arid conditions obtained in that region all through the Pliocene and Pleistocene, with tropical faunas undergoing decline during those times. Nevertheless, areas of greater humidity and more abundant vegetation persisted there, even during periods of glacial maxima in Europe (Tchernov, 1968). These surely provided ecological environments suitable for faunal expansion.

It has been shown for the Sahara (de Heinzelin, 1963; Quezel and Martin, 1962; Bakker, 1962, 1964) that extreme fluctuations in vegetation accompanied small changes in climate during the late Pleistocene, and open forests stood where desert conditions prevail today. It seems likely that climatic fluctuations of the kind prevailing during the late Quaternary to Recent would have occurred at earlier times as well, and would have allowed sufficient shifting of vegetation belts to permit the expansion of at least some faunal elements from Africa to Asia via the Middle East. Frenzel (1968) has shown that forest-steppe and steppe-desert conditions prevailed east of the Dead Sea, south of the Carpathians and across to Burma during the early Pliocene. The forest-steppe habitat could have provided a pathway for elephants from Africa to India during pre-Praetiglian times.

Elephas is first recorded in Asia as *E. planifrons* in the Tatrot horizon of the Siwalik Hills. The species rapidly expanded southward into Burma (upper Irrawaddies), Java (Tjidoelang, Djetis), and China (Shansi, Hsiachaohwan). *E. planifrons* became extinct before deposition of the Siwalik Boulder Conglomerate, presumably owing to competition with the more progressive *E. hysudricus*. It does not seem to have persisted in China beyond Hsiachaohwan time, being absent in the early Pleistocene deposits of Yunnan and Kiangsu. During the early Pleistocene, the species moved into the island of Celebes, clearly crossing a

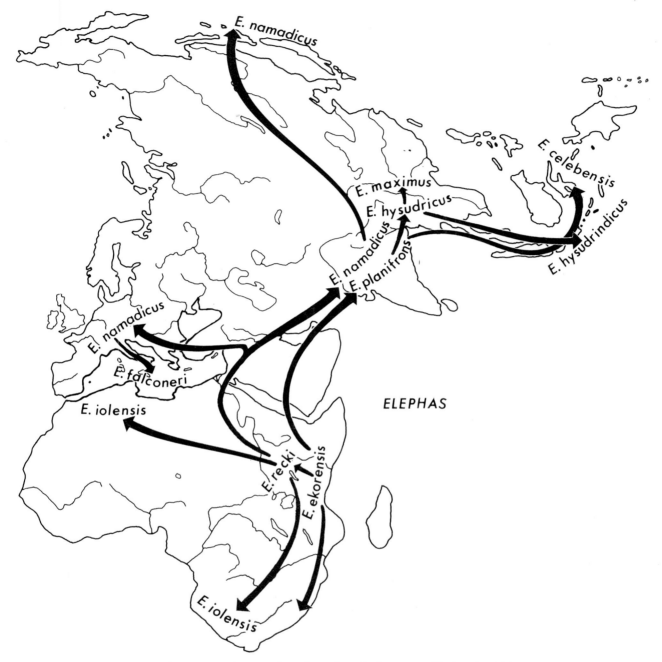

Fig. 48. Proposed expansion of species of the genus *Elephas* in Africa, Europe and Asia. Arrows are not intended to represent exact paths.

water barrier, and there underwent rapid dwarfing (fig. 48).

Elephas hysudricus first appears in the Pinjor horizon of India, probably derived from an early stage of *E. planifrons*, and quickly spread into Burma (upper Irrawaddies) and southern China (Kiangsu). It seems to have been short-lived in all areas, but a probable descendant, *E. hysudrindicus*, penetrated into Java in the (?) middle Pleistocene, where it is represented at Kedoeng and Sangiran. No trace of the transition between *E. hysudricus* and *E. maximus* is known, but there is no reason for believing that the latter originated anywhere but in southern Asia.

Late in the early Pleistocene, a second emigration from Africa seems best to account for *E. namadicus*. Although unknown in the Middle East (with the possible exception of the molar fragments from Ubeidiya in Israel) before the middle Pleistocene of Latemné, this

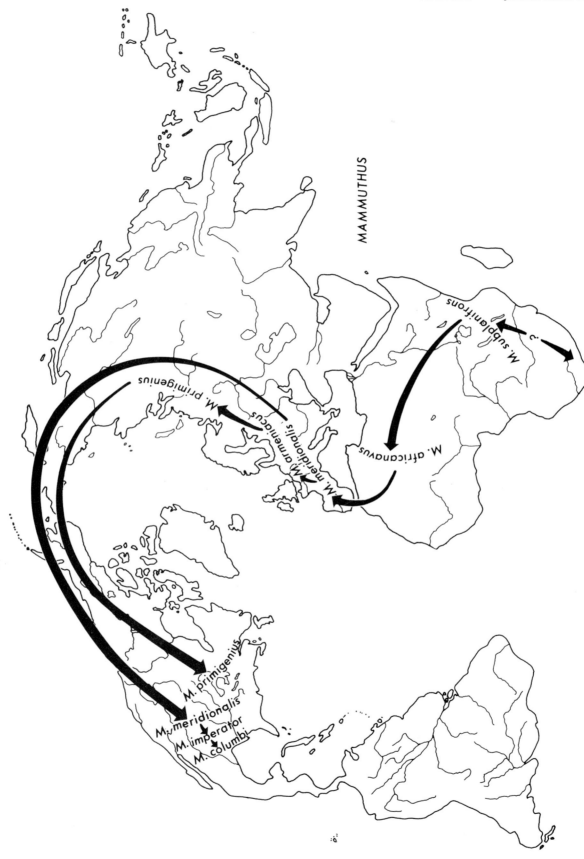

Fig. 49. Proposed geographic expansion of species of the genus *Mammuthus*. Arrows are not intended to represent exact paths.

line may have spread from there into both Europe and Asia during the latest phases of the early Pleistocene. *E. namadicus* as such quickly spread into India where it is well represented in middle Pleistocene deposits of the Narbadda and Godavari Valleys and the upper Karewa beds of Kashmir. The species rapidly expanded to dominance in Asia, occurring as the only elephant at this time in Java (Trinil, Ngandong), southern China (Szechuan and Yunnan), and Japan, persisting into the latest phases of the epoch at Sjara-osso-Gol in Inner Mongolia.

The same species appears in the European middle Pleistocene steppe faunas of Lower Mosbach and Süssenborn, and the temperate forest fauna of Mauer and the Cromer Forest Beds. During the warmest phases of the last two interglacials, the European and Asiatic populations may have been in direct contact via southern Siberia. During glacial maxima, however, the species' range would have contracted southward into the southern parts of Asia and Europe. It was probably during the Mindel low-water interval that the species moved onto the Sicily-Malta Bank where later isolation by rising sea levels led to the dwarfed populations of these islands. Sweepstakes crossings must have resulted in similar events on Crete and Cyprus at about the same time. These last three dwarf populations underwent parallel development in size reduction, presumably in response to insular pressures.

The genus *Mammuthus* originated as part of the early Pliocene radiation of the Elephantinae in Africa. Its earliest known stage, *M. subplanifrons*, quickly expanded into East and South Africa where it is represented in the roughly contemporaneous deposits at Langebaanweg, Chiwondo, Kaiso, and Kanam. By late Pliocene times, the more progressive species, *M. africanavus*, had spread northward into the Chad basin and was also represented in North Africa (Ichkeul, Fouarat, etc.). At about the same time the genus disappeared from sub-Saharan Africa and never re-expanded southward. The reason for this is not clear, but the local extinction of *Mammuthus* here coincides with the emergence of progressive populations of *Loxodonta* and *Elephas* in this area. The data are inconclusive for southern Africa, but several isolated molars may represent this species, in which case *Mammuthus* would have persisted there until the late Pliocene. In any event, there is no evidence to indicate that *Mammuthus* survived south of the Sahara beyond the closing phases of the Pliocene.

In the early Villafranchian, *Mammuthus* spread into southern Europe, possibly across the Betic-Riff Massif which seems to have persisted until this time (fig. 49). European mammoths are first recorded from scattered localities in Italy (Laiatico, Montopoli) and the Netherlands (Praetiglian Beds) as the Laiatico Stage of *M. meridionalis*. By the closing phase of the Pliocene the more progressive Montavarchi Stage had been attained,

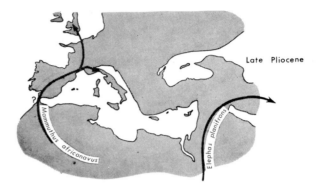
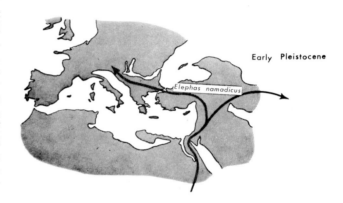
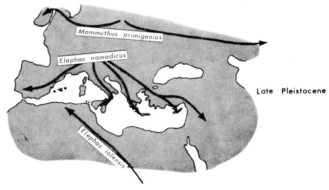

FIG. 50. Proposed Mediterranean corridors for elephantid expansion during the late Pliocene, early Pleistocene, and late Pleistocene. The species involved in each expansion are indicated.

and this form spread throughout Europe, crossing into England [20] where it is represented in the Red Crag deposits. The same form may have crossed the Mediterranean, this time probably by a cross-water route, for it is recorded at Ain Hanech in Algeria. (It is possible, however, that the Ain Hanech elephant represents a parallel line derived from a North African population of *M. africanavus*.)

[20] England was probably joined to continental Europe by a land connection during this time (Frenzel, 1968).

The mammoth lineage in Europe and northern Asia became progressively cold-adapted, evolving through successive grades (*M. armeniacus* and *M. primigenius*). The woolly mammoth survived into the latest Pleistocene of both Europe and Asia.

The *Mammuthus* group crossed the Bering Strait during the latest Villafranchian equivalent, and first appears in Irvingtonian faunas of Idaho, Kansas and Nebraska. Its earliest record is in the Bruneau fauna of Idaho (Malde and Powers, 1962) and has been dated at 1.36 m.y. (Evernden *et al.*, 1964). The successive forms *M. imperator* and *M. columbi* remained exclusively North American in distribution. During the late Pleistocene Rancholabrean stage, a second trans-Bering migration introduced the European woolly mammoth, *M. primigenius*, into North America where it was a common element in northern deposits of Wisconsin age (fig. 49).

We may briefly summarize the geographic dispersal patterns of the Elephantidae as follows (fig. 50).

Loxodonta—origin probably in sub-Saharan Africa, expanding throughout the continent by middle Pleistocene times, but becoming rare in East Africa until the latest Pleistocene. The genus never ranged outside of the continent.

Elephas—origin probably in sub-Saharan Africa with rapid expansion into eastern and southern areas, and only later ranging into North Africa. An early expansion into Asia via the Aden land-bridge in late Pliocene time resulted in a radiation of forms in the physiographically complex southern Asiatic region. A second expansion from Africa probably via the Middle East during the closing phases of the early Pleistocene moved from here into Europe and Asia during the middle Pleistocene.

Mammuthus—origin in sub-Saharan Africa with rapid expansion northward, becoming extinct in its former range. European access in the latest Pliocene was followed by expansion into North America late in the early Pleistocene. During the late Quaternary the woolly mammoth quickly spread throughout northern Asia and into North America.

X. SUMMARY

The proboscidean family Elephantidae represents the last major radiation within the order; it was characterized by a rapid adaptive shift and proliferation of species that dispersed into every part of the world except South America, Australia, and Antarctica. The history of the family covers a span of only about eight million years and exhibits a pattern of radiation and speciation that suggests external factors in its sudden extinction. The family was still actively dispersing and diversifying when, in the latest Pleistocene, extinction befell most of the lineages in Africa, Europe, Asia, and North America, with only two species surviving to the Recent epoch.

The family is divided into two subfamilies, the Stegotetrabelodontinae, which contains the earliest and least specialized forms, and the Elephantinae, which includes all later elephants. A total of twenty-five species are here considered valid for the family and these are placed in five generic groups as follows: *Stegotetrabelodon: S. syrticus, S. orbus; Primelephas: P. gomphotheroides, P. kororotensis; Loxodonta: L. adaurora, L. atlantica, L. africana; Elephas: E. ekorensis, E. recki, E. iolensis, E. namadicus, E. planifrons, E. celebensis, E. hysudricus, E. hysudrindicus, E. maximus, E. falconeri, E. platycephalus; Mammuthus: M. subplanifrons, M. africanavus, M. meridionalis, M. armeniacus, M. primigenius, M. imperator, M. columbi*.

The major evolutionary trends in the elephants seem to have centered around the restriction of the masticatory function to a fore-and-aft horizontal shearing mechanism and a freeing of the jaw musculature from compensation of weight-stress relationships not directly concerned with the functional requirements of the dentition. Thus, the jaw was shortened and the masticatory musculature was reorganized to eliminate rotational torque caused in other mammals by the forwardly placed center of gravity of the mandible. The result is a balanced lower dentition with muscles arranged roughly either radial or tangential to the curved occlusal surface formed by the molar teeth. Each muscle mass functions to give precise control of the shearing battery without the need to counterbalance forces created by other elements of the system. The unique architectural modifications of the elephant skull and jaw are functionally related to this adaptive strategy and are understandable only as parts of an integrated complex.

The analysis of evolutionary trends in the various elephant lineages suggests that *Loxodonta* specialized early in its history with respect to cranial structure, but that it remained relatively unspecialized in its dentition. The particular mode of life of this group, whatever this may have been, apparently placed few additional demands on the dental-cranial complex that was already established by early Pliocene time.

Mammuthus and especially *Elephas* showed a rapid initial specialization in dental adaptations, far surpassing *Loxodonta* in this respect, but did not undergo any marked cranial modification until well into the Pleistocene. These two groups expanded into Europe and Asia where competition with each other and with other proboscidean groups was probably great. Climatic fluctuations, at least in Europe, must also have provided intense selection pressures that affected many aspects of the ecology of these animals. Subsequent dental evolution was accompanied by rapid cranial specialization, the latter paralleling the course taken much earlier by *Loxodonta*. Thus, contrary to earlier belief, *Loxo-*

donta cannot be considered a persistently "primitive" line, but rather a highly efficient and specialized group whose primitive-looking teeth are an integral part of the total functional system that is well adapted to its particular ecological requirements.

The geographic distribution of the Elephantidae during the Pliocene and Pleistocene was largely controlled by geologic events in the Mediterranean basin that determined the corridors and sweepstakes routes available to elephants during this period. The ecological habitats of the species involved were also an important factor. Thus, *Loxodonta* never expanded beyond the African continent, probably because of its sub-Saharan distribution during the Pliocene/Pleistocene when routes were generally broadly open, and also because of climatic conditions in Europe during the later Pleistocene. *Elephas* dispersed out of Africa on two separate occasions. One early form entered Asia during the middle Pliocene and culminated in the living Asiatic elephant. Another, more specialized group spread into Europe and Asia during the latest Pliocene. *Mammuthus* managed a trans-Mediterranean crossing, probably via a Gibraltar land-bridge, during the late Pliocene, and rapidly spread throughout Europe, Asia, and, later, North America.

The rapid evolution of elephants during their entire history, and their remarkably wide geographic distribution, make them valuable for the correlation of Pliocene and Pleistocene deposits on a local, as well as on a world-wide scale. They also permit the study of evolutionary phenomena along several parallel lineages over a considerable time interval. Few other mammalian groups of Pleistocene age have so far proved as useful.

XI. ACKNOWLEDGMENTS

I am deeply indebted to the large number of people who contributed in many ways to the present study. Without their generous help this work could not have been undertaken, let alone completed.

Special gratitude is expressed to Professor Bryan Patterson who was responsible for my initial involvement in African paleontology and in this project. His helpful advice and perceptive criticisms throughout the period of this study were freely given and most welcome. For encouragement and useful discussions I thank Professor Ernest Williams. Special thanks are also due to Professor H. B. S. Cooke for his encouragement and valued advice.

For discussions on various aspects of paleontological, stratigraphical, functional, correlational, and evolutionary problems I wish to thank the following: E. Aguirre, L. Ambrosetti, the late C. Arambourg, A. Azzaroli, A. K. Behrensmeyer, A. Berzi, W. W. Bishop, C. K. Brain, J. Chavallion, R. Clark, H. B. S. Cooke, Y. Coppens, S. Coryndon, A. W. Crompton, G. Eck, F. Fitch, A. Gentry, A. J. F. Gogelein, J. Harris, J. de Heinzelin, Q. B. Hendey, A. Hill, D. A. Hooijer, F. C. Howell, E. N. Keen, B. Kurtén, L. S. B. Leakey, M. D. Leakey, R. E. Leakey, R. Leidy, J. Mawby, E. Mayr, J. A. Miller, A. Nanda, B. Patterson, T. Shikama, R. Singer, A. W. Walker, L. H. Wells, and R. C. Wood.

The following persons and institutions made possible my examination of specimens relevant to the present study: E. Aguirre, Department of Geology, University of Madrid; C. Arambourg, Muséum National d'Histoire Naturelle, Paris; F. Azzabi, Natural History Museum, Tripoli; A. M. Bailey, Denver Natural History Museum; T. Barry, South African Museum; G. Battetta, Muséum d'Histoire Naturelle, Lyon; A. Berzi, Museo di Geologia e Paleontologia, Universitá de Firenze; W. W. Bishop, Department of Geology, Bedford College; C. K. Brain, Transvaal Museum; Dr. Chatterjee, Geological Museum, Calcutta; K. H. Fisher, Institut für Paleontologie, Humboldt Universität, East Berlin; L. Gianelli, Museo de Paleontologia, University of Pisa; C. Guerin, Geology Department, University of Lyon; K. Hatai, Institute of Geology and Paleontology, Tôhoku University; A. C. Hoffman, National Museum, Bloemfontein; D. A. Hooijer, Rijksmuseum van Natuurlijke Historie; F. C. Howell, Anthropology Department, University of California, Berkeley; J. Hürzeler, Naturhistorisches Museum, Basel; G. L. Jepsen, Department of Geology, Princeton University; T. Kamei, Department of Geology and Mineralogy, Kyoto University; G. Kortenbout van der Sluys, Geological Museum, Leiden; L. S. B. Leakey, Centre for Prehistory and Paleontology, Nairobi; J. Lehman, Muséum National d'Histoire Naturelle, Paris; R. Liversidge, McGregor Memorial Museum; Dr. Petronio, Department of Geology, Cittá Universitaria, Rome; D. E. Savage, Department of Paleontology, University of California, Berkeley; C. B. Schultz, University of Nebraska State Museum; R. Singer, Department of Anatomy, University of Chicago; A. Sutcliffe, British Museum (Natural History); F. Takai, Geological Institute, Tokyo University; B. S. Tavari, Department of Geology, Panjab University; R. Tedford, American Museum of Natural History.

The following persons read part or all of the manuscript and offered helpful criticism: W. W. Bishop, A. W. Crompton, H. B. S. Cooke, B. Kurtén, R. Leidy, E. Mayr, B. Patterson, and E. E. Williams. Any errors that remain are, of course, solely my responsibility.

I would like to extend my appreciation and thanks to the Game Department, Republic of Kenya, for its help in the procurement of an African elephant for study, and to Dr. A. W. Walker, University College, Nairobi, for his cooperation and assistance in the dissection of the elephant head for the analysis of jaw function.

Dr. J. Walsh of the Department of Mines and Geology, Nairobi, provided valuable assistance in the preparation for field work in Kenya. A. D. Lewis and his associates prepared the many (and heavy) fossil elephants and other specimens from Kanapoi and Lothagam, and Miss Catherine McGeary typed the final manuscript. Mrs. F. Biondi skillfully prepared the frontispiece reconstruction.

For financial support of the field work and museum study programs, I gratefully acknowledge grants from the National Science Foundation (U.S.A.) (grants number G.A. 425 and 1188), the National Geographic Society, the Museum of Comparative Zoology, The Wenner-Gren Foundation for Anthropological Research, and an Evolutionary Biology Training Grant to the Department of Biology, Harvard University (N.S.F. grant number BG-7346).

XII. LITERATURE CITED

Adams, A. Leith. 1870. *Notes of a Naturalist in the Nile Valley and Malta* (Edinburgh).

——. 1881. *Monograph on British Fossil Elephants* (London).

Adams, M. 1808. "Some Account of a Journey to the Frozen Sea, and of the Discovery of the Remains of a Mammoth." *Phil. Mag. (Tilloch)* **29**: pp. 141–153.

Aguirre, E. 1969. "Evolutionary History of the Elephant." *Science* **164**: pp. 1366–1376.

Allen, G. M. 1936. "Zoological Results of the George Vanderbilt African Expedition of 1934. Part II—The Forest Elephant of Africa." *Proc. Acad. Nat. Sci. Phila.* **88**: pp. 15–44.

Ambrosetti, P. 1968. "The Pleistocene Dwarf Elephant of Spinagallo." *Geol. Rom.* **8**: pp. 277–366.

Arambourg, C. 1938. "Mammifères fossiles du Maroc." *Mém. Soc. Sci. Nat. Maroc* **46**: pp. 1–74.

——. 1947. "Contribution a l'étude géologique et paléontologique du bassin du Lac Rudolphe et de la basse vallée de l'Omo." *Miss. Scient. de l'Omo, 1932–1933* **2**: pp. 232–562.

——. 1949. "Les gisements de vertébrés villafranchiens de l'Afrique du Nord." *Bull. Soc. Géol. France,* ser. 5, **19**: pp. 195–203.

——. 1952. "Note préliminaire sur quelques éléphants fossiles de Berbérie." *Bull. Mus. nat. d'Hist. Nat., Paris,* ser 2, **24**: pp. 407–418.

——. 1969–1970. "Les Vertébrés du Pléistocène de l'Afrique du Nord." *Arch. Mus. nat. d'Hist. Nat., Paris,* ser. 7, **10**: pp. 1–126.

Athanasiu, S. C. 1912 (1915). "Resturile de Mammifère Pliocene superiora e dela Tuluçesti in districtul Covurlui." *Anuar. Inst. Geol. Romaneie* **6**: pp. 408–415.

Azzaroli, A. 1970. "Villafranchian Correlations Based on Large Mammals." *Giornale di Geol.,* ser. 2, **35**, 1: pp. 1–21.

Bakker, E. M. Van Zinderen. 1962. *Palynology in Africa.* 7th rept. Univ. O.F.S Bloemfontein.

——. 1964. *Palynology in Africa.* 8th rept. Univ. O.F.S., Bloemfontein.

Barbour, E. H. 1915. "A New Nebraska Mammoth, *Elephas hayi.*" *Amer. Jour. Sci.,* ser 4, **40**, 236: pp. 129–134.

Bate, D. M. S. 1903. "Preliminary Note on the Discovery of a Pigmy Elephant in the Pleistocene of Cyprus." *Proc. Roy. Soc. London* **71**, 475: pp. 498–500.

——. 1907. "On Elephant Remains from Crete, with Description of *Elephas creticus,* sp. n." *Proc. Zool. Soc. London,* pp. 238–250.

Behrensmeyer, A. K. 1970. "Preliminary Geological Interpretation of a New Hominid Site in the Lake Rudolf Basin." *Nature* (London), **226**, 5242: pp. 235–236.

Berggren, W. A. 1969. "Cenozoic Chronostratigraphy, Planktonic Foraminiferal Zonation and the Radiometric Time Scale." *Nature* (London) **224**: pp. 1072–1075.

Bishop, W. W. 1967. "The Later Tertiary in East Africa—Volcanics, Sediments and Faunal Inventory." *In: Background to Evolution in Africa,* W. W. Bishop and J. D. Clark, eds. (Chicago, University Chicago Press), pp. 31–56.

Bizard, C., A. Bonnet, J. Freulon, G. Gerard, A. F. de Lapparent, M. Lelubre, P. Vincent, and P. Wacrenier. 1955. "Sur l'extension de couches continentales tertiaires (Continental terminal) dans le nord-est du bassin du Tchad." *C. R. Acad. Sci., Paris* **241**, 24: pp. 1800–1803.

Blainville, H. M. D. de. 1845. *Ostéographie ou description iconographique comparée du squellette et du système dentaire des Mammifères. III. Quaternates* (Paris).

Blow, W. H. 1969. "Late Middle Eocene to Recent Planktonic Foraminiferal Biostratigraphy." *In: Proc. 1st. Int'l. Conf. Planktonic Microfossils, Geneva,* P. Bronniman and H. H. Renz, eds. *1967* **1**: pp. 199–421.

Blumenbach, J. H. 1797. *Handbuch der Naturgeschichte* (5th ed., Göttingen, J. H. Dietrich).

——. 1799. *Handbuch der Naturgeschichte* (6th ed., Göttingen, J. H. Dietrich).

——. 1803. *Manuel d'Histoire Naturelle, Traduit de l'Allemand, de J. Fr. Blumenbach, par Soulange Artoud* (2v, Metz).

Boule, M., and P. Teilhard de Chardin. 1928. "Paléontologie." *In:* M. H. Breuil, E. Licent and P. Teilhard de Chardin, "La paléolithique de la Chine," *Arch. Inst. Paléont. Humaine, Mem.* no. 4, part 2: pp. 27–102.

Brain, C. K. 1958. "The Transvaal Ape-man-bearing Cave Deposits." *Transvaal Mus. Mem.,* no. 11.

Brandt, J. F. 1833. "Ueber die Existenz von sechs Arten vorweltlicher Elephanten, die in Zahnban dem asiatischen Elephanten ahneln." *Mem. Acad. Imp. Sci. St. Petersb.,* ser. 6, **2**, Math. and Phys., Bull. Sci., no. 2: pp. x–xv.

Breyne, J. P. 1741. "A letter from John Phil. Breyen, M.D. F.R.S. to Sir Hans Sloane, Bart. Pres. R. S. with observations, and a description of some mammoth's bones dug up in Siberia, proving them to have belonged to elephants." *Phil. Trans. Roy. Soc. London* **40** (1737–1738): pp. 124–138.

Brown, F. H. and K. R. Lajoie. 1970. "Radiometric Age Determinations on Pliocene/Pleistocene Formations in the Lower Omo Basin, Ethiopia." *Nature* (London) **229**: pp. 483–485.

Burnett, G. T. 1830. "Illustrations of the Quadrupeda, or Quadrupeds, Being the Arrangement of the True Four-footed Beasts Indicated in Outline." *Quart. Jour. Sci.,* London (Dec. 1829): pp. 336–353.

Busk, G. 1867. "Description of the Remains of Three Extinct Species of Elephant, Collected by Capt. Spratt, C.B., R.N., in the Ossiferous Cavern of Zebbug, in the Island of Malta." *Trans. Zool. Soc. London* **6**, 5: pp. 227–306.

Colbert, E. H. 1940. "Pleistocene Mammals from the Ma Kai Valley of Northern Yunnan, China." *Amer. Mus. Novit.,* no. 1099: pp. 1–10.

——. 1943. "Pleistocene Vertebrates Collected in Burma." *Trans. Amer. Philos. Soc.* **32**: pp. 395–430.

Colbert, E. H., and D. A. Hooijer. 1954. "Pleistocene Mammals from the Limestone Fissure of Szechwan, China." *Amer. Mus. Nat. Hist., Bull.* **102**, 1: pp. 1–134.

Cooke, H. B. S. 1947. "Variation in the Molars of the Living African Elephant and a Critical Revision of the Fossil Proboscidea of Southern Africa." *Amer. Jour. Sci.* **245**: pp. 434–457; 492–517.

——. 1949. "Fossil Mammals of the Vaal River Deposits." *Union S. Afr. Geol. Surv., Mem.* no 35, III.

——. 1960. "Further Revision of the Fossil Elephantidae of Southern Africa." *Palaeont. Afric.* **7**: pp. 59–63.

——. 1964. "Pleistocene Mammal Faunas of Africa, with Particular Reference to Southern Africa." *In: African Ecology and Human Evolution,* F. C. Howell and F. Bourliere, eds. (Chicago, Aldine Pub. Co.), pp. 65–116.

——. 1967. "The Pleistocene Sequence in South Africa and Problems of Correlation." *In: Background to Evolution in Africa,* W. W. Bishop and J. D. Clark, eds. (Univ. Chicago Press), pp. 175–184.

Cooke, H. B. S., and S. Coryndon. 1970. "Fossil Mammals from the Kaiso Formation and Other Related Deposits in Uganda." *Fossil Vertebrates of Africa* **2**: pp. 107–224.

Cooke, H. B. S., and J. D. Clark. 1939. "New Fossil Elephant Remains from the Victoria Falls, Northern Rhodesia, and a Preliminary Note on the Geology and Archaeology of the Deposit." *Trans. Roy. Soc. S. Afr.* **27**: pp. 287–319.

COOKE, H. B. S., and R. F. EWER. 1972. "Fossil Suidae from Kanapoi and Lothagam, Northwestern Kenya." *Bull. Mus. Comp. Zool.* **143**, 3: pp. 149–295.

COOKE, H. B. S., and V. J. MAGLIO. 1972. "Plio/Pleistocene Stratigraphy in East Africa in Relation to Proboscidean and Suid Evolution." *In: Calibration of hominoid evolution,* W. W. Bishop and J. A. Miller, eds. 1971 Wenner-Gren Symposium (Scottish Academic Press), pp. 303–329.

COPE, E. D. 1898. Syllabus of lectures on the Vertebrata. Philadelphia, Univ. Penn.

COPPENS, Y. 1965. "Les Proboscidiens du Tchad." *Actes du Ve Congr. Panafr. de Préhist. et de l'étude de Quat. (Santa Cruz de Tenerife),* pp. 331–387.

COX, A. 1969. "Geomagnetic Reversals." *Science* **163**: pp. 237–245.

CRUSAFONT-PAIRÓ, M., and S. REGUANT. 1970. "The Nomenclature of Intermediate Forms." *Syst. Zool.* **19**: pp. 254–257.

CUVIER, G. 1798. *Tableau Élémentaire de l'Histoire Naturelle des Animaux* (Paris).

———. 1799. "Mémoire sur les Espèces d'Éléphans Vivantes et Fossiles." *Mém. Inst. Nat. Sci. et Arts,* Sci., Mathém. et Phys., II, Fructidor, an VII (1799): pp. 1–22.

———. 1806. "Sur les Éléphans Vivans et Fossiles." *Ann. Mus. nat. d'Hist. Nat., Paris* **8**: pp. 1–58; 93–155; 249–269.

———. 1825. *Recherches sur les Ossimens Fossiles de Quadrupèdes, où l'on Rétablit les caractères de Plusieurs Espèces d'Animaux dont les Révolutions du Globe ont Détruit les Espèces* (3rd ed., Paris).

CUVIER, F., and É. GEOFFROY SAINT-HILAIRE. 1825. *Histoire Naturelle des Mammifères avec des Figures originales, dessinées d'Après des Animaux vivans etc.,* III. Livr. LI, LII, 1825.

DART, R. 1927. "Mammoths and Man in the Transvaal." *Nature Supplement,* no. 3032: pp. 41–48.

———. 1929. "Mammoths and Other Fossil Elephants of the Vaal and Limpopo Watersheds." *S. Afr. Jour. Sci.* **26**: pp. 698–731.

DE HEINZELIN, J. 1963. "Observations of the Absolute Chronology of Upper Pleistocene." *In: African ecology and human evolution,* F. C. Howell and F. Bourliere, eds. (Chicago, Aldine Press), pp. 295–303.

DE KAY, J. E. 1842. *Natural History of New York.* Part I, Zoology: Zoology of New York, or the New York Fauna.

DEPÉRET, C., and L. MAYET. 1923. "Les Éléphants Pliocènes. Part II." *Ann. Univ. Lyon.,* n.s. I, **42**, II: pp. 89–213.

DERANIYAGALA, P. E. P. 1955. "Some Extinct Elephants, Their Relatives and the Two Living Species. *Ceylon Nat. Mus. Pub.* (Columbo).

DIEN, M. N., and L. P. CHIA. 1938. "Cave and Rock Shelter Deposits in Yunnan." *Geol. Soc. China, Bull.* **18**: pp. 326–349.

DIETRICH, W. O. 1916. "*Elephas antiquus Recki* n.f. aus dem Diluvium Deutsch-Ostafrikas." *Arch. Biontol.* **4**, 1: pp. 1–80.

———. 1942. "Ältesquärtare Säugetiere aus de sudlichen Serengeti, Deutsch-Ostafrikas." *Palaeontogr.* **94**, A: pp. 43–133.

———. 1958. "Übergangsformen des Südelefanten (*Elephas meridionalis* Nesti) in altpleistozän Thüringens." *Geologie* **7**: pp. 797–807.

———. 1965. "Fossile Elephantenzähne von Voigtstedt Thüringen." *Paläont. Abh. Abt. A,* **2**, 3/4: pp. 521–536.

DUBOIS, E. 1891. "Voorloopig bericht omtrent het onderzoek naar de Pleistocene en Tertiaire Vertebraten-fauna van Sumatra en Java, gedurende het jaar 1890." *Natuurk. Tijdschr. Ned. Indie* **51**: pp. 93–100.

———. 1908. Das geologische Alter der Kendeng-Oder Trinil fauna. *Tijdschr. Ned. Aardr. Genoots, Amsterdam,* ser. 2, **25B**, 6: pp. 1235–1270.

EICHWALD, E. 1835. "De Pecorum et Pachydermorum reliquis fossilibus in Lithuania, Volhynia et Podolia Repertis." *Nova Acta Acad. Leop. Carol.* **17**: pp. 677–760.

ERICSON, D. B., and G. WOLLIN. 1968. "Pleistocene Climates and Chronology in Deep Sea Sediments." *Science,* **162**: pp. 1227–1234.

EVERNDEN, J. F., D. E. SAVAGE, G. H. CURTIS and G. J. JAMES. 1954. "Potassium-Argon Dates and the Cenozoic Mammalian Chronology of North America." *Amer. Jour. Sci.* **262**: pp. 145–198.

EWER, R. F. 1957. "Faunal Evidence on the Dating of the Australopithecinae." *Proc. 3rd Panafr. Congr. Prehist. (Livingston, 1955):* pp. 135–142.

———. 1967. "The Fossil Hyaenids of Africa—a Reappraisal. *In: Background to Evolution in Africa,* W. W. Bishop and J. D. Clark, eds. (Univ. Chicago Press), pp. 109–123.

FALCONER, H. 1857. "On the Species of Mastodon and Elephant Occurring in the Fossil State in Great Britain. Part I. Mastodon." *Quart. Jour. Geol. Soc. London* **13**: pp. 307–360.

———. 1865. "On the Species of Mastodon and Elephant Occurring in the Fossil State in Great Britain. Part II. Elephant." *Quart. Jour. Geol. Soc. London* **21**: pp. 253–332.

———. 1862. "On Ossiferous Caves in Malta, Explored by Captain Spratt, R.N., C.B., with an Account of *Elephas meletensis,* a Pigmy Species of Fossil Elephant, and Other Remains Found in Them." *Parthenon* **1**, 25: p. 780.

———. 1863. "On the American Fossil Elephant of the Regions Bordering the Gulf of Mexico (*Elephas columbi,* Falc.), with General Observations on the Living and Extinct Species." *Nat. Hist. Rev.* **3**: pp. 43–114.

———. 1868a. *Description of the Plates of the Fauna Antiquua Sivalensis, from Notes and Memoranda,* Charles Murchison, ed. (London, Hardwicke).

———. 1868b. *Palaeontological Memoires and notes of the late Hugh Falconer ... with a biographical sketch of the author,* Charles Murchison, ed. (2 v., London).

FALCONER, H., and P. T. CAUTLEY. 1845–1849. *Fauna Antiquua Sivalensis, Being the Fossil Zoology of the Sewalik Hills in the North of India.* Figures only. 1945: Part I; 1946: Part II; 1947: Parts III-VIII; 1949: Part IX (London, Smith, Elder and Co.).

———. 1846. *Fauna Antiquua Sivalensis.* Letterpress. Part I (London, Smith, Elder and Co.).

FEJFAR, O. 1969. "Die Nager aus den Kiesen von Süssenborn bei Weimar." *Paläont. Abh. Abt. A,* **3**, 2/3: pp. 761–770.

FERNEX, F., C. LORENY and J. MAGNÉ. 1965. "A propos de l'âge de la mise-en-place des nappes betiques (Espagne méridionale)." *C.R. Acad. Sci. Paris* **260**: pp. 933–936.

FISCHER DE WALDHEIM, G. 1829a. "Notice sur Quelques Animaux Fossiles de la Russie." *Nouv. Mém. Soc. Imp. Nat. Moscou* **I**: pp. 283–299.

———. 1829b. "Fossiles. Notice sur le Mammont." *Bull. Soc. Imp. Nat. Moscou,* Première Année, no. 9: pp. 267–278.

FITCH, F. J., and J. A. MILLER. 1970. "Radiometric Age Determinations of Lake Rudolf Artifact Site." *Nature* (London) **225**, 5242: pp. 236–238.

FLEROV, K. K. 1969. "Die Bison-Reste aus den Kiesen von Süssenborn bei Weimar." *Paläont. Abh. Abt. A,* **3**, 3/4: pp. 489–520.

FRANZ, H. 1967. "On the Stratigraphy and Evolution of Climate in the Chad Basin During the Quaternary." *In: Background to evolution in Africa,* W. W. Bishop and J. D. Clark, eds. (Univ. Chicago Press), pp. 273–284.

FRENZEL, B. 1968. "The Pleistocene Vegetation of Northern Eurasia." *Science* **161**: pp. 637–649.

GARUTT, V. E. 1954. "Novye dannye o drevneishikh slonakh. Rod *Protelephas* gen. nov. (New data on the earliest elephants, genus *Protelephas* gen. nov.)." *Doklady An SSSR (Earth Sciences)* **114**, 1: pp. 189–191.

———. 1957. "O novom Iskopaemom slone *Phanogoroloxodon mammontoides* gen. et sp. nov. S Kaukaza. (A new extinct elephant, *Phanogoroloxodon mammontoides* gen. et sp. nov., from the Caucasus)." *Doklady An SSSR* **112**, 2: pp. 333–335.

GAUDRY, A. 1888. *Les Ancêtres de nos Animaux dans les Temps Géologiques* (Paris).

GEOFFROY SAINT-HILAIRE, É. 1837. "Encore cet écrit sur le Sivatherium, considéré sous le point de vue de ses révélations contestées relativement à la Philosophie zoologique." *C.R. Acad. Sci., Paris* **4**, 4: pp. 113–122.

GIERMANN, G. 1962. "Meeresterrassen am Nordufer de Strasse von Gibraltar." *Bei. Naturf. Ges. Freiburg i. Br.* **52**: pp. 111–118.

GIGNOUX, M. 1955. *Stratigraphic Geology* (San Francisco, W. H. Freeman and Co.).

GOLDFUSS, G. A. 1823. "Osteologische Beiträge zur Kenntness Verschiedener Saugetiere der Vorwelt." *Nova Acta Acad. Leop. Carol.* **11**: pp. 451–490.

GRAELLS, D. M. DE LA PAY. 1897. "Fauna Mastodológia Ibérica." *Mem. Real. Acad. Cien. Exactas, Fis. nat., Madrid* **17**: pp. 1–806.

GRAY, J. E. 1821. "On the Natural Arrangement of Vertebrate Animals." *London Medical Repository* **15**, 88: pp. 296–310.

GREGORY, W. K. 1903. "Adaptive Significance of the Shortening of the Elephant's Skull." *Amer. Mus. Nat. Hist., Bull.* **19**: pp. 387–394.

GUENTHER, E. W. 1969. "Die Elephanten molaren aus den Kiesen von Süssenborn bei Weimar." *Paläont. Abh. Abt. A,* **3**, 3/4: pp. 711–734.

HAAS, G. 1963. "Preliminary Remarks on the Early Quaternary Faunal Assemblage from Tel Ubeidiya, Jordan Valley." *S. Afr. Jour. Sci.* **59**: pp. 73–76.

———. 1966. "On the Vertebrate Fauna of the Lower Pleistocene Site 'Ubeidiya." *Jerusalem, the Israel Acad. Sci. and Hum.*

HAMMON, A. H. 1970. "Deep Sea Drilling: A Giant Step in Geological Research." *Science* **170**: pp. 520–521.

HAUGHTON, S. H. 1932. "On Some South African Proboscidea." *Trans. Roy. Soc. S. Afr.* **21**: pp. 1–18.

HAY, O. P. 1922. "Further Observations on Some Extinct Elephants." *Proc. Biol. Soc. Wash.* **35**: pp. 97–102.

HAY, O. P., and H. J. COOK. 1928. "Preliminary Descriptions of Fossil Mammals Recently Discovered in Oklahoma, Texas and New Mexico." *Proc. Colo. Mus. Nat. Hist.* **8**, 2, Part I: p. 33.

———. 1930. "Fossil Vertebrates Collected near, or in Association with, Human Artifacts at Localities near Colorado, Texas; Frederick, Oklahoma; and Folsom, New Mexico." *Proc. Colo. Mus Nat. Hist.* **9**, 2: pp. 4–40.

HEINTZ, E. 1967. "Données Préliminaires sur les Cervidés Villafranchiens de France et d'Espagne." *Coll. Intern. de Centre Nat. Rech. Scientifique, N.* 163: "Problèmes actuels de Paléontologie" (Evolution des Vertébrés), (Paris), pp. 539–552.

HENDEY, Q. B. 1969. "Quaternary Fossil Sites in the Southwestern Cape Province." *S. Afr. Archaeol. Bull.* **24**, 3 and 4: pp. 95–105.

———. 1970. "The Age of the Fossiliferous Deposits at Langebaanweg, Cape Province." *Ann. S. Afr. Mus.* **56**, 3: pp. 119–131.

HIBBARD, C. W. 1953. "*Equus (Asinus) calobatus* Troxell and Associated Vertebrates from the Pleistocene of Kansas." *Trans. Kansas Acad. Sci.* **56**, 1: pp. 111–126.

HIBBARD, C. W., D. E. RAY, D. E. SAVAGE, D. W. TAYLOR and J. E. GUILDAY. 1965. "Quaternary Mammals of North America." *In: The Quaternary of the United States, VII Congr. Int'l Ass. Quat. Res.* (Princeton Univ. Press), pp. 509–525.

HOOIJER, D. A. 1949. "Pleistocene Vertebrates from Celebes. IV. *Archidiskodon celebensis* nov. spec." *Zool. Med. Mus. Leiden* **30**, 14: pp. 205–226.

———. 1952. "Fossil Mammals and the Plio-Pleistocene Boundary in Java." *Proc. Kon. Ned. Akad. v. Wetensch. Amsterdam* ser. B, **55**: pp. 436–443.

———. 1953a. "Pleistocene Vertebrates from Celebes. V. Lower Molars of *Archidiskodon celebensis* Hooijer." *Zool. Med. Mus. Leiden* **31**, 28: pp. 311–318.

———. 1953b. "Pleistocene Vertebrates from Celebes. VII. Milk Molars and Premolars of *Archidiskodon celebensis* Hooijer." *Zool. Med. Mus. Leiden* **32**, 20: pp. 221–230.

———. 1954a. "Pleistocene Vertebrates from Celebes. XI. Molars and a Tusked Mandible of *Archidiskodon celebensis* Hooijer." *Zool. Med. Mus. Leiden* **33**, 15: pp. 103–120.

———. 1954b. "A Pygmy *Stegodon* from the Middle Pleistocene of Eastern Java." *Zool. Med. Mus. Leiden* **33**, 14: pp. 91–102.

———. 1955a. "*Archidiskodon planifrons* (Falconer and Cautley) from the Tatrot Zone of the Upper Siwaliks." *Leidse Geol. Med.* **20**: pp. 110–119.

———. 1955b. "Fossil Proboscidea from the Malay Archipelago and the Punjab." *Zool. Verband. Mus. Leiden*, No. 28: pp. 1–146.

———. 1956a. "The Lower Boundary of the Pleistocene in Java and Age of *Pithecanthropus.*" *Quaternaria* **3**: pp. 5–10.

———. 1956b. "*Epileptobos* gen. nov. for *Leptobos groeneveldtii* Dubois from the Middle Pleistocene of Java." *Zool. Med. Mus. Leiden* **34**, 17: pp. 239–241.

———. 1958. "An Early Pleistocene Mammalian Fauna from Bethlehem." *Brit. Mus. (Nat. Hist.), Bull., Geol.* **3**, 8: pp. 267–292.

———. 1959. "Fossil Mammals from Jisr Banat Yaqub, South of Lake Huleh, Israel." *Res. Coun. Israel, Bull.* **8G**, 4: pp. 177–199.

———. 1961. "Middle Pleistocene Mammals from Latemné, Orontes Valley, Syria." *Ann. Archeol. de Syrie* **11**: pp. 117–132.

———. 1964. "New Records of Mammals from the Middle Pleistocene of Sangiran, Central Java." *Zool. Med. Mus. Leiden* **40**, 10: pp. 73–88.

———. 1969. "Pleistocene East African Rhinoceroses." *Fossil Vertebrates of Africa* **7**: pp. 71–98.

HOOIJER, D. A., and B. PATTERSON. 1972. "Rhinoceroses from the Pliocene of Northwestern Kenya." *Bull. Mus. Comp. Zool.* **144**, 1: pp. 1–26.

HOPWOOD, A. T. 1926. "The Geology and Paleontology of the Kaiso Bone-beds. Part II. Paleontology." *Geol. Surv. Uganda, Occ. Pap.* no. 2: pp. 13–36.

———. 1939. "The Mammalian Fossils." *In:* T. P. O'Brien, *The Pre-history of Uganda Protectorate* (Cambridge).

HOWELL, F. C. 1959. "The Villafranchian and Human Origins." *Science* **130**: pp. 831–844.

———. 1960. European and Northwest African Middle Pleistocene Hominids. *Curr. Anthrop.* **1**, 3: pp. 195–232.

———. 1967. "Later Studies in Africa and Paleoanthropology: A Post-conference Appraisal. *In: Background to Evolution in Africa*, W. W. Bishop and J. D. Clark, eds. (Univ. Chicago Press), pp. 903–922.

HÜRZELER, J. 1966. "Nouvelles Découvertes de Mammifères dans les sédiments fluvio-Lacustres de Villafrancha d'Asti." Prob. Act. Paléont., *Coll. Int. du Cent. Nat. de la Rech. Scient.*, no. 163: pp. 133–136.

IKEBE, N. 1969. "A Synoptical Table on the Quaternary Stratigraphy of Japan." *Osaka City Univ., Jour. Geosci.* **12**: pp. 45–51.

JANOSSY, D. 1965. "Die Insectivoren-Reste aus dem Altpleistozän von Voigtstedt in Thüringen." *Paläont. Abh. Abt. A,* **2**, 2/3: pp. 665–678.

JEPSEN, G. L. 1943. "Time, Strata and Fossils: Comments and Recommendations." *In: Time and Stratigraphy in the Evolution of Man*, C. B. Hunt, W. L. Straus and M. G. Wolman, eds., *Nat. Acad. Sci., Pub.* no. 1469: pp. 88–97.

JOURDAN, C. 1861. "Desterrains sidérolitiques." *C. R. Acad. Sci. Paris* **53**: pp. 1009-1014.

KALKE, H. D. 1961. "Revision der Säugetierfauna der Klassischen deutchen Pleistozän-Fundstellen von Süssenborn, Mosbach und Taubach." *Géologie* **10**: pp. 493–596.

———. 1965a. "Die Cerviden-Reste aus den Tonen von Voigtstedt in Thüringen." *Paläont. Abh. Abt. A*, **2**, 2/3: pp. 381–426.

———. 1965b. "Die Rhinocerotiden-Reste aus den Tonen von Voigtstedt in Thüringen." *Ibid.*, pp. 453–519.

———. 1965c. "Die stratigraphische Stellung de Faunen von Voigtstedt." *Ibid.*, pp. 691–692.

———. 1969a. "Die *Ovibos*-Reste aus den Kiesen von Süssenborn bei Weimar." *Paläont. Abh. Abt. A*, **3**, 3/4: pp. 521–529.

———. 1969b. "Die *Soergelia*-Reste aus den Kiesen von Süssenborn bei Weimar." *Ibid.*, pp. 531–545.

———. 1969c. "Die Cerviden-Reste aus den Kiesen von Süssenborn bei Weimar." *Ibid.*, pp. 547–610.

———. 1969d. "Die Rhinocerotiden-Reste aus den Kiesen von Süssenborn bei Weimar." *Ibid.*, pp. 667–709.

KENT, P. E. 1941. "The Recent History and Pleistocene Deposits of the Plateau North of Lake Eyasi, Tanganyika." *Geol. Mag. (London)* **78**: pp. 173–184.

KRETZOI, M. 1950. "*Stegoloxodon* nov. gen., a loxodonta elefantok estleges azsiai ose." *Foldtani Kozlony* **80**: pp. 405–408.

———. 1965a. "Die Nager und Lagomorphen von Voigtstedt in Thüringen und ihre chronologische Aussage." *Paläont. Abh. Abt. A*, **2**, 2/3: pp. 587–660.

———. 1965b. "Die Amphibiens aus der altpleistozänen Funstellen Voigtstedt in Thüringen." *Ibid.*, pp. 325–333.

KUMMEL, B. 1970. *History of the Earth* (2nd ed., San Francisco, Freeman and Co.).

KURTÉN, B. 1957. "Mammal Migrations, Cenozoic Stratigraphy and the Age of the Australopithecines." *Jour. Paleont.* **31**, 1: pp. 215–227.

———. 1959. "The Age of the Australopithecinae. *Acta Univ. Stockholmensis, Stokholm Contr. Geol.* **6**, 2: pp. 9–22.

———. 1963. "Villafranchian Faunal Evolution." *Soc. Sci. Fennica, Comment. Biol.* **26**, 3: pp. 3–18

———. 1968. *Pleistocene Mammals of Europe* (Chicago, Aldine Pub. Co.).

LEAKEY, L. S. B. 1965. *Olduvai Gorge 1951-1961.* Vol. I: Fauna and background (Cambridge Univ. Press).

LEAKEY, R. 1969. "New Cercopithecidae from the Chemeron Beds, Kenya." *Fossil Vertebrates of Africa* **1**: pp. 53–69.

———. 1970. "Fauna and Artifacts from a New Plio-Pleistocene Locality near Lake Rudolf in Kenya. *Nature* (London) **226**, 5242: pp. 223–224.

LEHMAN, U. 1953. "Eine Villafranchiano-Fauna von der Erpfinger Höhle (Schwabische Alb)." *Neues Jb. Geol. Paläont., Mh.*, pp. 437-464.

———. 1957. "Weitere Fossilfunde aus der ältesten Pleistozän der Erpfinger Höhle (Schwabische Alb)." *Mitt. Geol. Staatsinst. Hamburg* **26**: pp. 60–99.

LEIDY, J. 1858. "Notice of Remains of Extinct Vertebrata, from the Valley of the Niobrara River, collected during the exploration expedition of 1857, in Nebraska, under the command of Lieut. G. K. Warren, U. S. Top. Eng., by F. V. Hayden, geologist to the expedition." *Proc. Acad. Nat. Sci. Phila.* **10**: pp. 20–29.

LESSON, R. P. 1842. *Nouveau Tableau du Règne Animal. Mammifères* (Paris).

LINK, H. F. 1807. Beschriebung Naturalien-Sammlung der Universität zu Rostock (4) 1807, 3 (cited by Osborn: 1942, p. 1384).

LINNAEUS, C. 1754. *Animalia Rariora, Imprimis et Exotica. Mus. Adolph. Frid. Regis, Holmiae* (Stockholm).

———. 1758. *Systema naturae per regina tria naturae, secundum classes, ordines, genera, species cum characteribus, differentus; synonymis, locis* (Editio decima, reformata, I. Laurentii Slavii, Holmiae).

LUDOLF, H. W. 1696. *Grammatica Russica* (Oxonii).

LYDEKKER, R. 1886. *Catalogue of the fossil Mammalia in the British Museum (Natural History)*, Part IV (London).

———. 1907. "Notes on Two African Mammals." *Proc. Zool Soc. London,* **2**; pp. 782-785.

MAAREL, F. H. VAN DER. 1932 "Contribution to the Knowledge of the Fossil Mammalian Fauna of Java." *Wet. Med. Dienst. Mijnb. Ned. Indie,* no. 15: pp. 1–208.

MACINNES, D. G. 1942. "Miocene and Post-Miocene Proboscidea from East Africa." *Trans. Zool. Soc. London* **25**: pp. 33–106.

MAGLIO, V. J. 1969. "The Status of the East African Elephant *Archidiskodon exoptatus* Dietrich 1942." *Breviora,* no. 336: pp. 1–25.

———. 1970a. "Early Elephantidae of Africa and a Tentative Correlation of African Plio-Pleistocene Deposits." *Nature* (London) **225**: pp. 328–332.

———. 1970b. "Four New Species of Elephantidae from the Plio-Pleistocene of Northwestern Kenya." *Breviora,* no. 341: pp. 1–43.

———. 1971. "The Vertebrate Faunas from the Kubi Algi, Koobi Fora and Ileret areas, East Rudolf, Kenya. *Nature* (London), **230**, 5000: pp. 248–249.

———. 1972a. "Vertebrate Faunas and Chronology of Hominid-bearing Sediments East of Lake Rudolf, Kenya." *Nature* (London) **239**: pp. 379–385.

———. 1972b. "Evolution of mastication in the Elephantidae." *Evolution* **26**, 4: pp. 638–658.

MAGLIO, V. J., and Q. B. HENDEY. 1970. "New Evidence Relating to the Supposed Stegolophodont Ancestry of the Elephantidae." *S. Afr. Archaeol. Bull.* **25**, 3-4: pp. 85–87.

MAJOR, C. I. FORSYTH. 1875. "Considerasioni sulla Fauna dei Mammiferi Pliocenici e post-Pliocenici della Toscana." *Atti. Soc. Toscana Sci. Nat.* **I**, 1: pp. 7–40; **I**, 3: pp. 223–245.

———. 1883. "Die Tyrrhenis." *Kosmos,* XIII, Jahrg. **7**: pp. 1–17.

MAKIYAMA, J. 1924. "Notes on a Fossil Elephant from Sahamma, Tôtômi." *Mem. Coll. Sci. Kyoto Imp. Univ.,* ser. B, **1**, 2: pp. 255–264.

MALDE, H. E., and H. A. POWERS. 1962. "Upper Cenozoic Stratigraphy of Western Snake River Plain, Idaho." *Geol. Soc. Amer. Bull.* **73**: pp. 1197–1220.

MARTIN, J. K. L. 1887. Fossil Säugetierreste von Java und Japan. *Samml. Geol. Reichsmus. Leiden, Beiträge Geol. Ost-Asiens u. Australiens,* ser. 1, **4**, 3: pp. 25–69.

———. 1888. "Neue Wirbeltierreste von Pati-Ajain auf Java." *Ibid.*, pp. 87–116.

MARTIN, P. S., and H. E. WRIGHT, JR. 1967. *Pleistocene Extinctions. The Search for a Cause* (New Haven, Yale University Press).

MARTYN, J, and P. V. TOBIAS. 1967. "Pleistocene Deposits and New Fossil Localities in Kenya." *Nature* (London), **215**: pp. 476–480.

MATHER, W. W. 1838. "Remarks in Addition to and Explanation of the Review of the Report of the Geological Survey of Ohio in a Letter to the Editor." *Amer. Jour. Sci.,* ser. 1, **34**: pp. 362–364.

MATHEWS, G. M., and T. IREDALE. 1912. "Perry's Arcana—An Overlooked Work." *The Victorian Naturalist*: the Jour. and Mag. of the Field Naturalists' Club of Victoria, Melbourne, **29**: pp. 7–16.

MATSCHIE, P. 1900. "Uber geographische Abarten des Afrikanischen Elephantens." *Sitz.-Ber. Ges. naturf. Freunde Berlin,* no. 8: pp. 189–197.

MATSUMOTO, H. 1924. "Preliminary Note on Fossil Elephants in Japan." *Jour. Geol. Soc. Tokyo* **31**, 371: pp. 255–272.

———. 1927. "On *Leith-Adamsia siwalikiensis*, a New Generic and Specific Name of Archetypal Elephants." *Japanese Jour. Geol. and Geogr.* **5**, 4: p. 213.

MATTAUER, M. 1963. "Le style tectonique des chaines tellienne et Rifaine." *Geol. Rundschau* **53**: pp. 296–313.

MAYR, E. 1963. *Animal Species and Evolution* (Cambridge, Harvard Univ. Press).

———. 1970. *Principles of Systematic Zoology* (New York, McGraw Hill).

MEIRING, A. J. D. 1955. Fossil Proboscidean Teeth and Ulna from Virginia, O.F.S." *Res. Nas. Mus. Bloemfontein* **1**, 8: pp. 187–201.

MORREN, C. F. A. 1834. *Mémoire sur les Ossemens Fossiles d'Éléphans trouvés en Belgique* (Ghent).

MUSIL, R. 1968. "Die Mammutmolaren von Předmosti (CSSR)." *Paläont. Abh. Abt. A*, **3**, 1: pp. 5–191.

NESTI, F. 1825. "Sulla nuove species de elephante fossile del Valdarno all 'Illustrissimo sig. Dott. Prof. Ottaviano Targioni Tozzetti (Lettere sopra alcune ossa fossili de Valdarno non per anco descritte)." *Nuov. Giorn. Lett.* **11**, 24: pp. 195–216.

OSBORN, H. F. 1921. "The Evolution, Phylogeny and Classification of the Proboscidea." *Amer. Mus. Novit.*, no. 1: pp. 1–15.

———. 1922. "Species of American Pleistocene Mammoths, *Elephas jeffersonii*, new species." *Amer. Mus. Novit.*, no. 41: pp. 1–16.

———. 1923. "New Subfamily, Generic and Specific stages in the Evolution of the Proboscidea." *Amer. Mus. Novit.*, no. 99: pp. 1–4.

———. 1924. "*Parelephas* in Relation to Phyla and Genera of the Family Elephantidae." *Amer. Mus. Novit.*, no. 152: pp. 1–7.

———. 1928. "Mammoths and Man in the Transvaal." *Nature* (London) **71**, 3052: pp. 672–673.

———. 1929. "New Eurasiatic and American Proboscideans." *Amer. Mus. Novit.* no. 393: pp. 1–22.

———. 1931. "*Palaeoloxodon antiquus italicus* sp. nov., Final Stage in the "*Elephas antiquus* Phylum." *Amer. Mus. Novit.*, no. 460: pp. 1–24.

———. 1932. "The '*Elephas meridionalis*' Stage Arrives in America." *Proc. Colo. Mus. Nat. Hist.* **11**, 1: pp. 1–3.

———. 1934. "Primitive *Archidiskodon* and *Palaeoloxodon* of South Africa." *Amer. Mus. Novit.*, no. 741: pp. 1–15.

———. 1936. *Proboscidea*. Vol. I. (New York, American Museum Press), pp. 1–802.

———. 1942. *Proboscidea*. Vol. II, pp. 805–1676.

PATERSON, T. T. 1941. "On a World Correlation of the Pleistocene." *Trans. Roy. Soc. Edinburgh*, **60**, 2: pp. 373–425.

PATTERSON, B., A. K. BEHRENSMEYER and W. D. SILL. 1970. "Geology and Fauna of a New Pliocene Locality in Northwestern Kenya." *Nature (London)* **226**, 5249: pp. 918–921.

PAVLOW, M. 1910. "Les Éléphants Fossiles de la Russie." *Nouveaux Mem. Imp. des Natural de Moscou* **27**, 2: pp. 1–56.

PEI, W. C. 1936. "The Mammalian Remains from Locality 3 at Choukoutien." *Pal. Sinica, C*, **7**, 5: pp. 1–108.

———. 1939. "An Attempted Correlation of Quaternary Geology, Paleontology and Prehistory in Europe and China." *Inst. Arch., Univ. London, Occ. Pap.* no. 2: pp. 1–16.

———. 1940. "The Upper Cave Fauna of Choukoutien." *Pal. Sinica*, n.s. *C*, **10**: pp. 1–84.

PERRY, G. 1811. *Arcana: or the Museum of Natural History: containing the most recent discovered objects. Embellished with coloured plates, and corresponding descriptions; with extracts relating to animals, and remarks of celebrated travellers, combining a general survey of nature* (London, G. Smeeton).

PETROCCHI, C. 1941. "I giacimento fossilifero di Sahabi." *Boll. Soc. Geol. Italiana* **60**, 1: pp. 107–114.

———. 1943. "I giacimento fossilifero di Sahabi." *Coll. Scient. Docum. a Cura (Min. A. I., IX)*, **12**.

———. 1953–1954. "I proboscidati di Sahabi.'" *Rendiconti Acad. Naz. Dei XL*, ser. 4, **4–5** (76–77 della fondazione): pp. 1–74.

PICCARD, L. 1937. "Inferences on the Problem of the Pleistocene Climate of Palestine and Syria Drawn from Flora, Fauna and Stratigraphy." *Proc. Prehist. Soc.*, n.s., **8**, 5: pp. 58–70.

PILGRIM, G. E. 1905. "On the Occurrence of *Elephas antiquus* (*namadicus*) in the Godavari Alluvium, with Remarks on the Species, its distribution and the Age of the Associated Indian Deposits." *Rec. Geol. Surv. India* **32**, 3: pp. 199–218.

———. 1910. "Preliminary Note on a Revised Classification of the Tertiary Freshwater Deposits of India." *Rec. Geol. Surv. India,* **40**, 3: pp. 185–205.

POHLIG, H. 1888. "Dentition und Kraniologie des *Elephas antiquus* Falc. mit Beiträgen uber *Elephas primigenius* Blum. und *Elephas meridionalis* Nesti, I." *Nova Acta Leop. Carol.* **53**, 1: pp. 1–279.

———. 1891. "Dentition und Kraniologie des *Elephas antiquus* Falc. mit Beiträgen uber *Elephas primigenius* Blum. und *Elephas meridionalis* Nesti, II." *Nova Acta Leop. Carol.* **57**, 5: pp. 267–466.

———. 1893. "Eine Elephanten hohle Siciliens und der erste Nachweis des cranialdomes von *Elephas antiquus*." *Abh. bayer. Akad.*, IIcl. **48**, 1: pp. 73–108.

POMEL, A. 1879. "Ossements d'Éléphants et d'Hippopotames découvertes dans une station préhistorique de la plaine d'Eghis (Province d'Oran)." *Bull. Soc. Géol. France*, ser. 3, **7**: pp. 44–51.

———. 1895. "Paléontologie Monographies, no. 6. Les Éléphants Quaternaires." *Carte Géol. l'Algérie*.

QUEZEL, P., and C. MARTINEZ. 1962. "Premiers résultats de l'analyse palynologique de sédiments recueillis au Sahara meridional à l'occasion de la mission Berliet Ténère-Tchad." *Extr. Doc. Sc.*, pp. 313–330.

RAMACCIONI, G. 1936. "L'*Elephas planifrons* de Laiatico (Pisa)." *Palaeont. Ital.*, n.s., **36**: pp. 215–233.

ROOSEVELT, T., and E. HELLER. 1914. *Life Histories of African Game Animals* (2v., London, J. Murray).

RUDDIMAN, W. F. 1971. "Pleistocene Sedimentation in the Equatorial Atlantic: Stratigraphy and Faunal Paleoclimatology." *Geol. Soc. Amer., Bull.*, **82**, 2: pp. 283–301.

SAVAGE, D. C., and G. H. CURTIS. 1970. "The Villafranchian Stage-age and Its Radiometric Dating." *Geol. Soc. Amer., Sp. Pap.* No. 124: pp. 207–231.

SAVAGE, R. J. G. 1967. "Early Miocene Mammal Faunas of the Tethyan Region." *In*: C. G. Adams and D. V. Ager, eds., *Aspects of Tethyan Biogeography*, Systematic Association Pub. no. 7: pp. 247–282.

SCHLOTHEIM, E. F. VON. 1820. *Die Petrefactenkunde auf ihrem jetzigen Standpunkte durch die Beschreibung seiner Sammlung versteinerter und fossiler Uberreste des Thiersund-Pflanzenreichs der Vorwelt erläutert* (Gotha).

SCOTT, W. B. 1907. "A Collection of Fossil Mammals from the Coast of Zululand." *Geol. Surv. Natal and Zululand*, 3rd and final rept., pp. 253–262.

SIMPSON, G. G. 1945. "The Principles of Classification and a Classification of Mammals." *Amer. Mus. Nat. Hist., Bull.* **85**: pp. 1–350.

———. 1961. *Principles of Animal Taxonomy* (New York and London, Columbia Univ. Press).

SINGER, R., and D. A. HOOIJER. 1958. "A *Stegolophodon* from South Africa." *Nature* (London), **182**: pp. 101–102.

SOERGEL, W. 1912. "*Elephas trogontherii* Pohlig und *Elephas antiquus* Falconer ihre Stammesgeschichte und ihre Bedeutung fur die Gliederung des deutschen Diluviums." *Palaeontogr.* **60**: pp. 1–114.

STAMP, L. D. 1922. "An Outline of the Tertiary Geology of Burma." *Geol. Mag.* **59**, 1: pp. 481–501.

STEFĂNESCU, S. 1924. "Sur la Présence de l'*Éléphas planifrons* et de trois mutations de l'*Éléphas antiquus* dans les couches géologiques de Roumanie." *C. R. Acad. Sci. Paris* **179**: pp. 1418–1419.

STOCK, C. 1935. "Exiled Elephants of the Channel Islands, California." *Sci. Monthly*, Sept.: pp. 205–214.

STOCK, C., and E. L. FURLONG. 1928. "The Pleistocene Elephants of Santa Rosa Island, California." *Science*, n.s., **58**, 1754: pp. 140–141.

STROSS, F. H. 1971. "Application of the Physical Sciences to Archeology." *Science* **171**, 3973: pp. 831–836.

SWARTZ, D. H., and D. D. ARDEN. 1960. "Geologic History of the Red Sea Area." *Bull. Amer. Assoc. Pet. Geol.* **44**, 10: pp. 1621–1637.

TCHERNOV, E. 1968. *Succession of Rodent Faunas During the Upper Pleistocene of Israel* (Mammalia Depicta, Hamburg and Berlin).

TEILHARD DE CHARDIN, P. 1936. "Fossil Mammals from Locality 9 of Choukoutien." *Pal. Sinica, C,* **7**, 4: pp. 1–61.

———. 1938. "The Fossils from Locality 12 of Choukoutien." *Pal. Sinica,* n.s., *C*, **5**: pp. 1–47.

TERRA, H. DE. 1939. "The Quaternary Terrace System of Southern Asia and the Age of Man." *The Geogr. Rev.* **29**, 1: 101–118.

TEMMINCK, C. J. 1847. *Coup d'Œil général sur les possessions Néerlandaises dans l'Inde Archipélagique* (3v., Leiden).

THENIUS, E. 1965. Ein Primaten-Rest aus dem altpleistozän von Voigtstedt in Thüringen. *Palaont. Abh. Abt. A,* **2**, 2/3: pp. 683–689.

THOMAS, G. 1956. "The Species Conflict." In: P. C. Sylvester-Bradley, ed., *The Species Concept in Paleontology,* The Systematics Association, London, Pub. no. **2**: pp. 17–32.

TILESIUS VON TILNAU, W. G. 1815. "De skeleto Mammonteo Sibirico ad moris glascialis littora anno 1807, effosso, Cui Praemissae Elephantini Generis Specierum Distinctiones." *Mém. Acad. Imp. Soc. St. Pétersb.,* ser. 5, **5**: pp. 406–513.

TOBIAS, P. V. 1966. "Fossil Hominid Remains from Ubeidiya, Israel." *Nature* **211**, 5045: pp. 130–133.

TOKUNAGA, S. 1934. "Fossil Teeth Found at Yokohama and Kakio, Kangawa Prefecture." *Jour. Geogr.* **46**, 546: pp. 363–371.

VAN BEMMELEN, R. W. 1950. *The Geology of Indonesia.* I. The Hague (Nijhoff).

VAN COUVERING, J. A. 1972. "Radiometric Calibration of the European Neogene." *In:* W. W. Bishop and J. A. Miller, eds., *Calibration of Hominoid Evolution* (Scottish Academic Press), pp. 247–271.

VAN VALEN, L. 1969. "Evolution of Communities and Late Pleistocene Extinctions." *Proc. N. Amer. Paleont. Conv. Part. E:* pp. 469–485.

VERRI, A. 1886. "Azione delle Forze nell'Assetto delle Valli con appendice sulla Distribuzione de Fossili nella Valdichiana e nell'Unbria Interna Settentrionale." *Boll. Soc. Geol. Ital.* **5**: pp. 416–454.

VONDRA, C. F., G. D. JOHNSON, B. E. BOWEN and A. K. BEHRENSMYER. 1971. "Preliminary Stratigraphic Studies of the East Rudolf Basin, Kenya." *Nature* (London) **231**: pp. 245–248.

VON KOENIGSWALD, G. H. R. 1934. "Zur Stratigraphie des javanischen Pleistocän." *De Ing. in Ned. Indie* **1**, 4: pp. 185–201.

———. 1935. "Die fossilen Säugetierfaunas Javas." *Proc. Kon. Akad. v. Wet. Amsterdam* **38**: pp. 188–198.

———. 1939. "Das Pleistocän Javas." *Quärtar (Berlin)* **2**: pp. 28–53.

———. 1950. "Vertebrate Stratigraphy." *In: The Geology of Indonesia,* Vol. I, General geology, R. W. van Bemmelen, ed. (The Hague, Nijhoff), pp. 91–93.

———. 1951. "Ein Elephant der *planifrons* Gruppe aus dem Pliocàen West-Javas." *Eclogae Geol. Helvetiae* **43**: pp. 268–274.

———. 1955. "Remarks on the Correlation of Mammalian Faunas of Java and India and the Plio-Pleistocene Boundary." *Kon. Ned. Akad. Wet. Amsterdam,* ser B, **5**, 3: pp. 204–210.

———. 1962. "Das absolute Alter des *Pithecanthropus erectus* Dubois." *In:* G. Kurth, ed., *Evolution und hominization,* pp. 112–119.

VON MEYER, H. 1832. *Palaeologica zur Geschichte der Erde und Geschopfe* (Frankfurt a. M.).

VON MEYER, H., and H. R. GIOPPERT. 1848. *Index palaeontologicus in Bronn's Handbuch einer Geschichte de Natur* **3**, 1.

WEITHOFER, K. A. 1888. "Einige Bemerkungen über der Proboscidier." *Morph. Jahrb.* **14**: pp. 507–516.

———. 1889. "Ueber die Tertiären Landsäugetiere Italiens." *Jahrb. Geol. Reichsanst.* **39**, 1: pp. 55–82.

———. 1890. "Die fossilen Proboscidier des Arnotales in Toskana." *Beitr. Pal. Osterr.-Ungarns* **8**: pp. 107–268.

ZEUNER, F. E. 1959. *The Pleistocene Period* (London, Hutchison).

ZUFFARDI, P. 1913. "Elephanti fossili del Piedmonte." *Palaeont. Ital.* **19**: pp. 121–187.

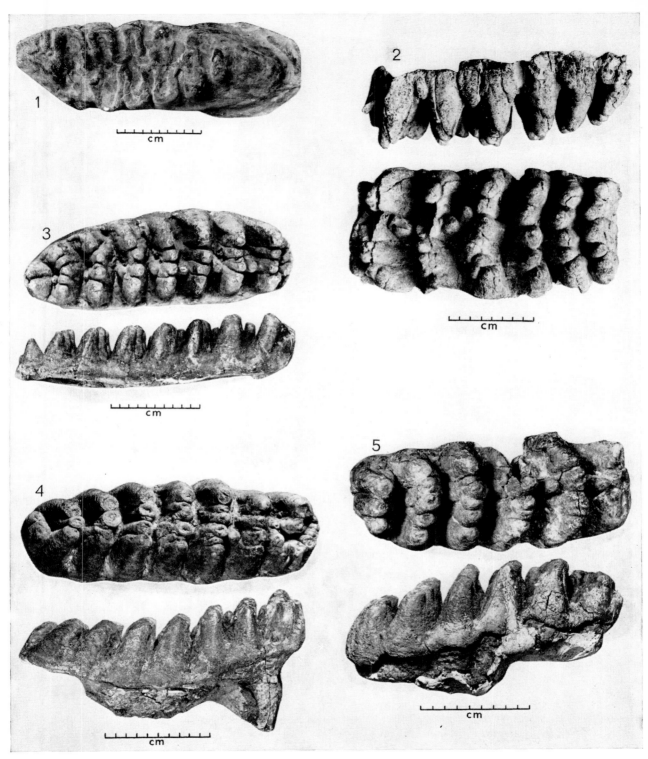

PLATE I

Stegotetrabelodon syrticus Petrocchi

1. MNHT (no number), *Type*, left M_3; cast. Sahabi, Libya.
2. IGR (no number), right M^3. Sahabi, Libya.
3. IGR (no number), right M_3. Sahabi, Libya.

Stegotetrabelodon orbus Maglio

4. KNM LT 359, right M_3. Lothagam 1, Kenya.
5. KNM LT 359, left M^3. Lothagam 1, Kenya.

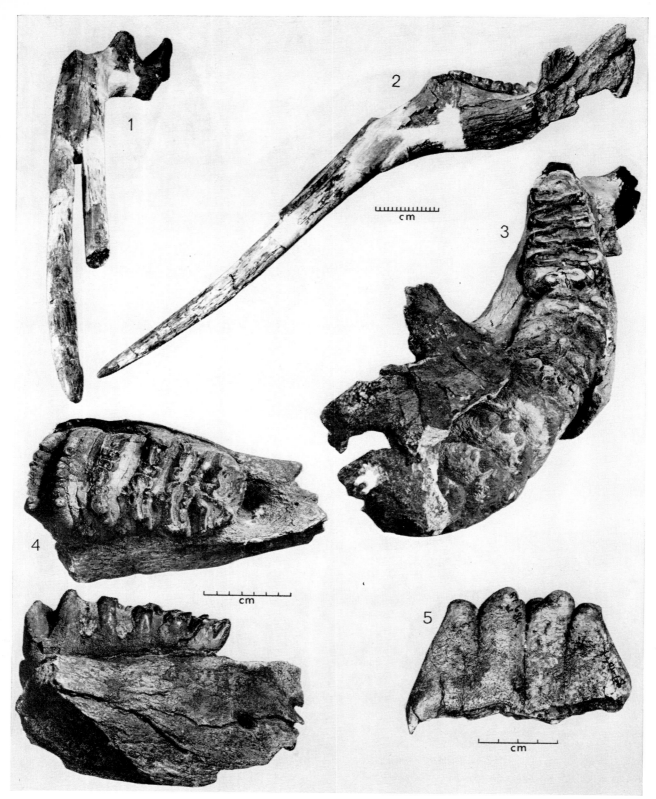

PLATE II

Stegotetrabelodon orbus Maglio

1. KNM LT 354, *Type*, mandible with M_2-M_3; anterior view. Lothagam 1, Kenya.
2. Same, left-lateral view.
3. Same, occlusal view.
4. KNM LT 342, right mandibular fragment with M_2. Lothagam 1, Kenya.
5. KNM LT 347, isolated molar plate of M^3. Lothagam 1, Kenya.

128

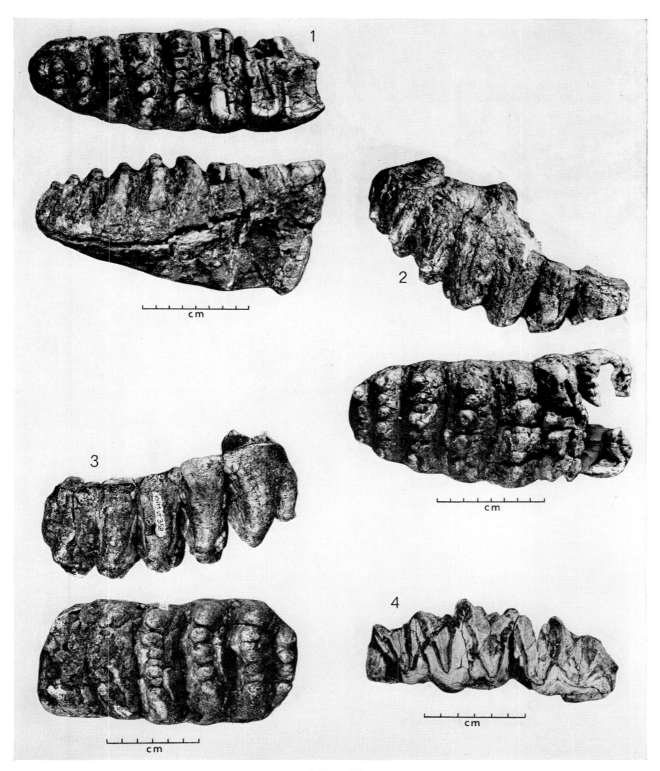

PLATE III

Primelephas gomphotheroides Maglio

1. KNM LT 351, *Type*, left M₃. Lothagam 1, Kenya.
2. KNM LT 351, *Type*, left M³. Lothagam 1, Kenya.
3. KNM LT 358, right M². Lothagam 1, Kenya.
4. KNM LT 351, *Type*, right M³; sectioned. Lothagam 1, Kenya.

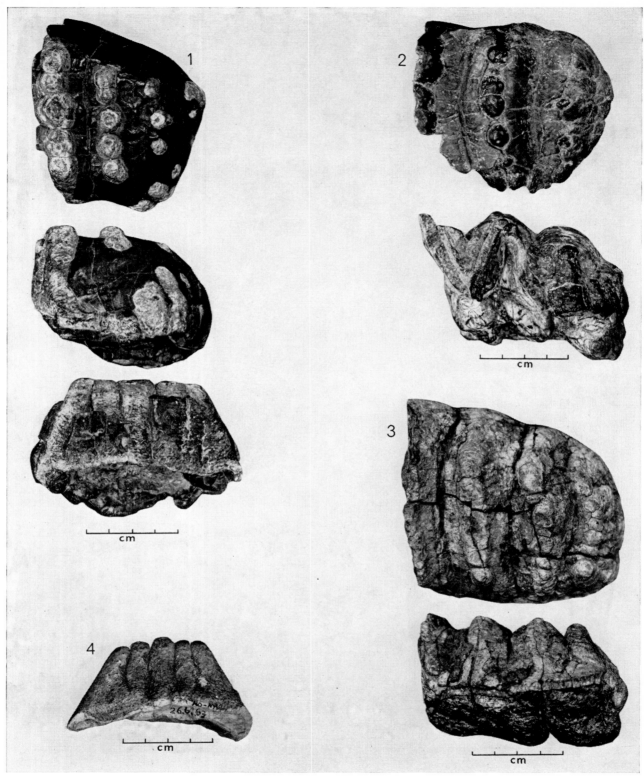

PLATE IV

Primelephas korotorensis (Coppens)
1. MNHN K-2-60, *Type*, right M_3. Koulá, Chad.
2. MNHN (no number), right M^3. Kolinga, Chad.

Primelephas cf. *gomphotheroides*
3. BM(NH) M-25160, right M^3. Nyawiega, Kaiso Formation, Uganda.
4. BM(NH) (no number), isolated plate. Nyawiega, Kaiso Formation, Uganda.

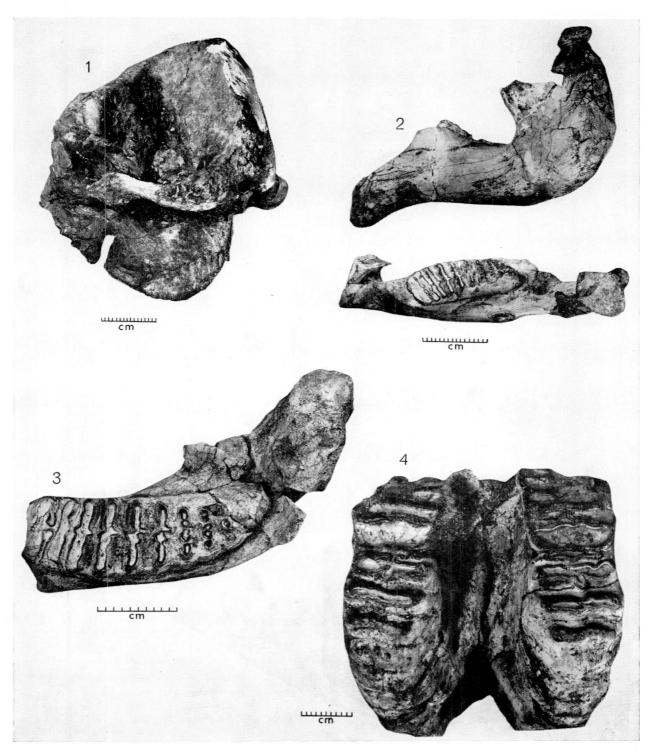

PLATE V

Loxodonta adaurora Maglio
1. KNM LT 353, skull with M^2-M^3. Lothagam 3, Kenya.
2. KNM KP 385, *Type*, left mandible with M_3. Kanapoi, Kenya.
3. KNM KP 385, *Type*, right mandible with M_3. Kanapoi, Kenya.
4. KNM ER 346, palate with left and right M^2-M^3. Kubi Algi, East Rudolf, Kenya.

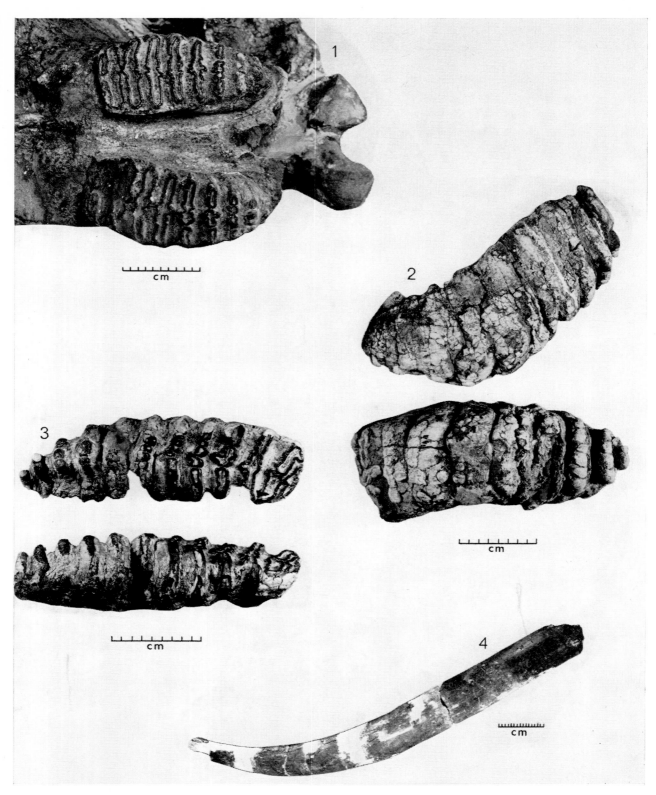

PLATE VI

Loxodonta adaurora Maglio

1. KNM KP 385, *Type*, palate of skull with left and right M^3. Kanapoi, Kenya.
2. KNM KP 383, left M^3. Kanapoi, Kenya.
3. KNM KP 407, right M_3. Kanapoi, Kenya.
4. KNM LT 353, left tusk. Lothagam 3, Kenya.

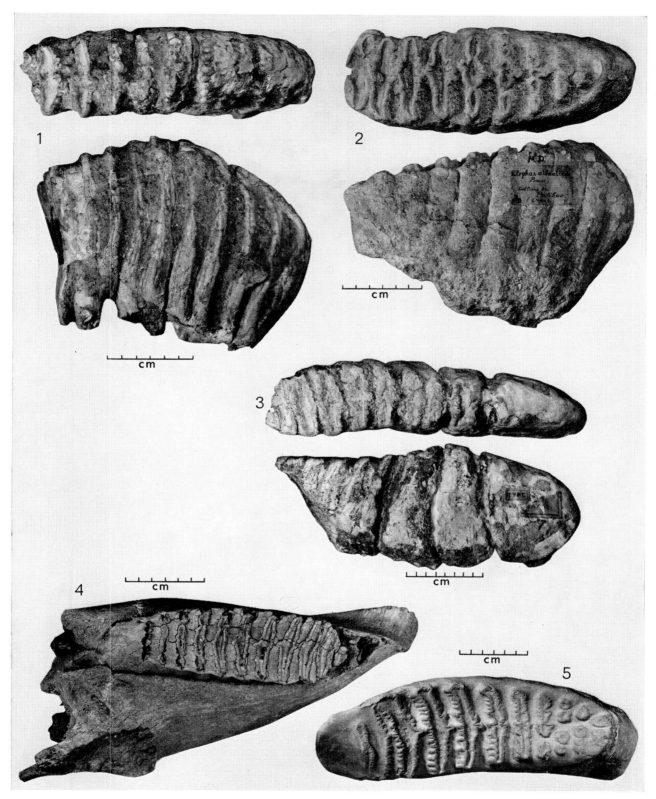

PLATE VII

Loxodonta atlantica (Pomel)

1. MNHN (no number), right M^2. Sablière de Palikao (Oran), Algeria.
2. MNHN (no number), left M^2. Ternifine.
3. MNHN (no number), left M_3. Ternifine.
4. SAM 2577, right M_3. Elandsfontein.
5. PU 11548, left M_3. Zululand.

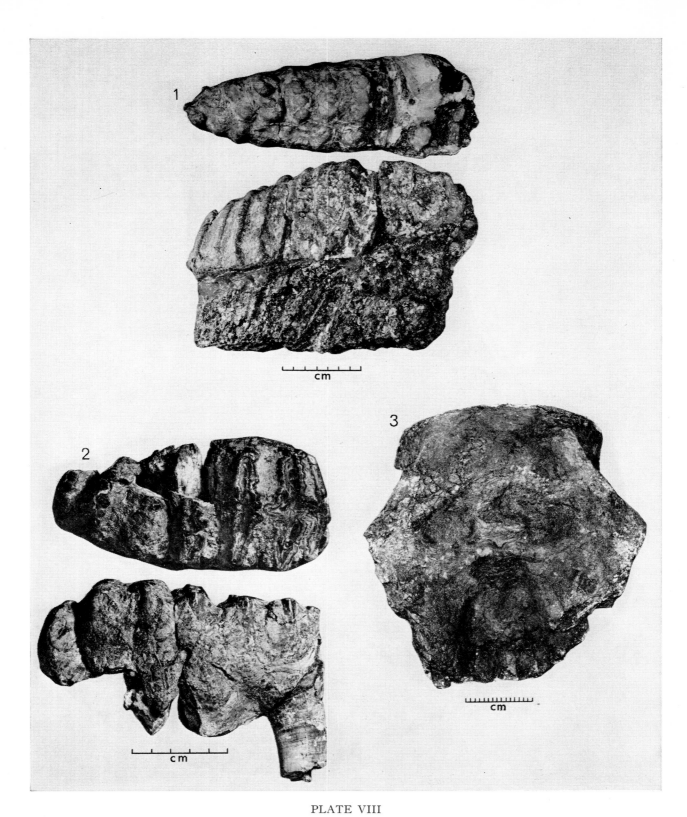

PLATE VIII

Elephas ekorensis Maglio
1. KNM EK 424, *Type*, right M^3. Ekora, Kenya.
2. KNM EK 420, left M^2. Ekora, Kenya.
3. KNM EK 422, skull. Ekora, Kenya.

PLATE IX

Elephas recki (Dietrich)

1. BM(NH) M-15211, left M_3. Stage 1. Kikagati, Uganda.
2. KNM ER 342, left M^3. Stage 2. Koobi Fora, East Rudolf, Kenya.
3. KNM L 26-45, left M_3. Stage 2. Shungura Formation, Ethiopia.
4. BM(NH) M-14691, left M^1. Stage 3. Olduvai Gorge, Bed I, Tanzania.
5. BM(NH) M-14694, left M^3. Stage 4. Olduvai Gorge, Bed IV, Tanzania.
6. KNM (no number), left M_3. Stage 4. Olorgesailie, Kenya.

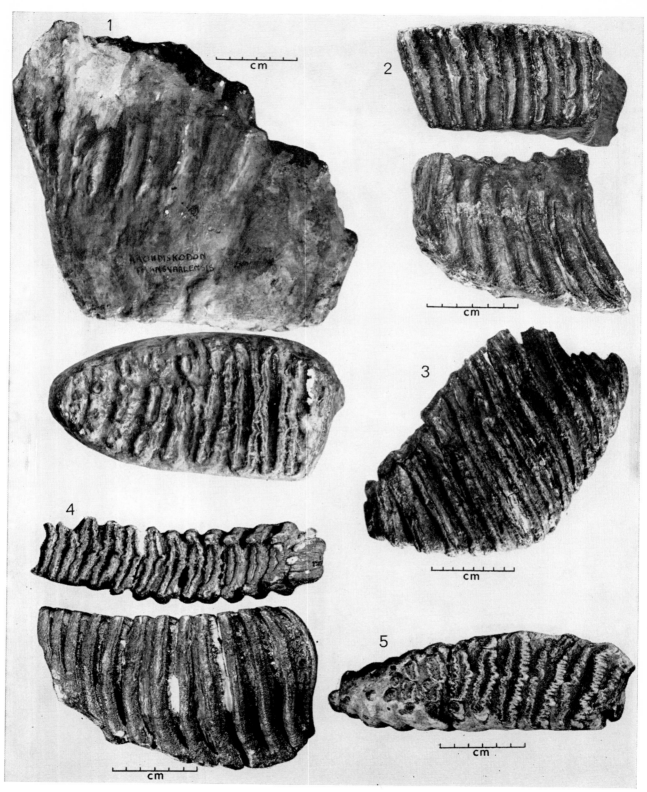

PLATE X

Elephas iolensis Pomel

1. MMK 4157, cast, right M^3. Bloemhof, South Africa.
2. MMK 4144, left M_1. Vaal River, Pniel, South Africa.
3. MMK 3930, right M^3. Vaal River, Delport's Hope, South Africa.
4. MNHN (no number), *Type*, left M_2. Beausejour Farm, Algeria.
5. MNHN (no number), right M_3. Port de Mastaganem, Algeria.

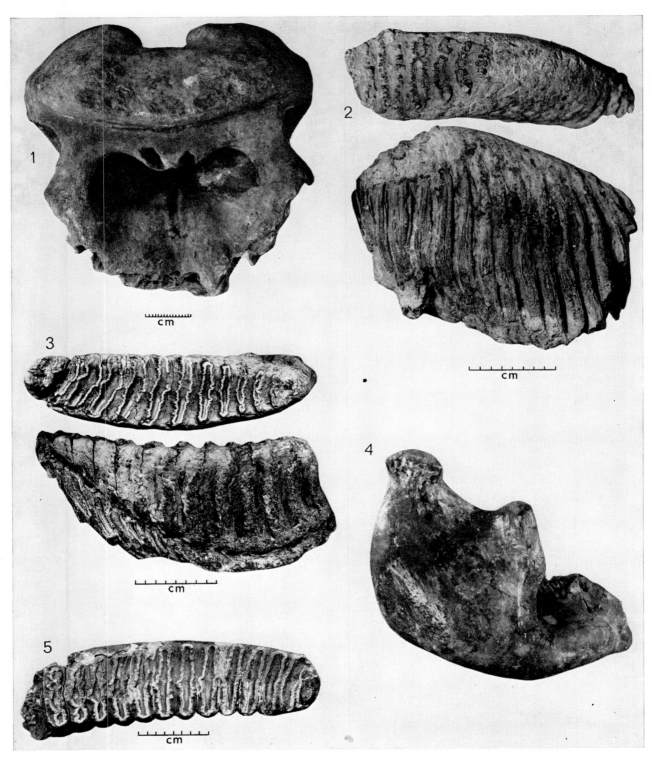

PLATE XI

Elephas namadicus Falconer and Cautley

1. BM(NH) M-3092, *Lectotype*, skull. Narbadda Valley, India.
2. NMB Ch. 699, left M^3. Montione, Italy.
3. NHB Ch. 495, right M_3. Chiantals, Italy.
4. NHB (no number), mandible.
5. NHB Sl. 1, right M_3. Blanzac-Solilhac (Ste. Loire), France.

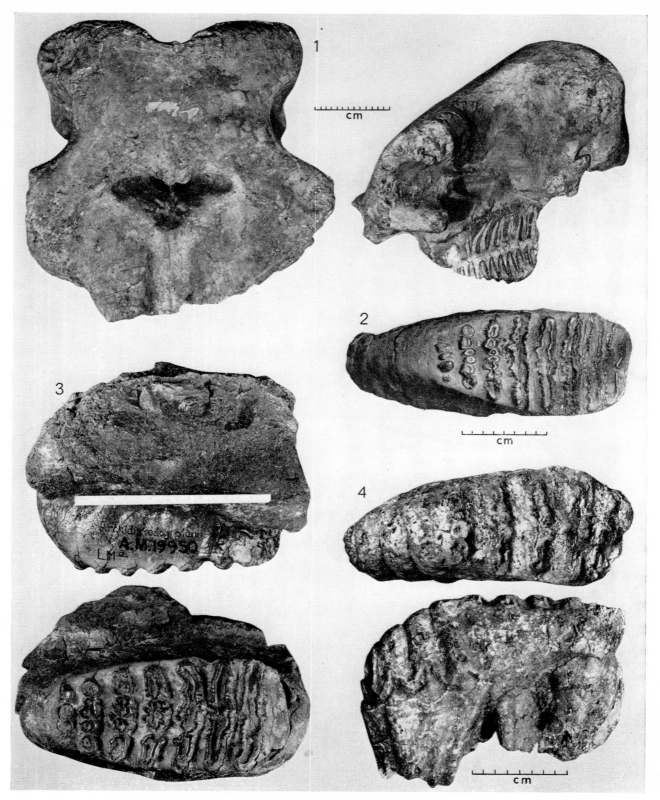

PLATE XII

Elephas planifrons Falconer and Cautley

1. BM(NH) M-3060, *Lectotype,* skull with left and right M^3. Pinjor horizon, Siwalik Hills, India.
2. BM(NH) M-3726, right M^3. Siwalik Hills, India.
3. AMNH 19950, left M^3. Chandigarh, Siwalik Hills, India.
4. BM(NH) M-18523, right M^3. Bethlehem, Jordan.

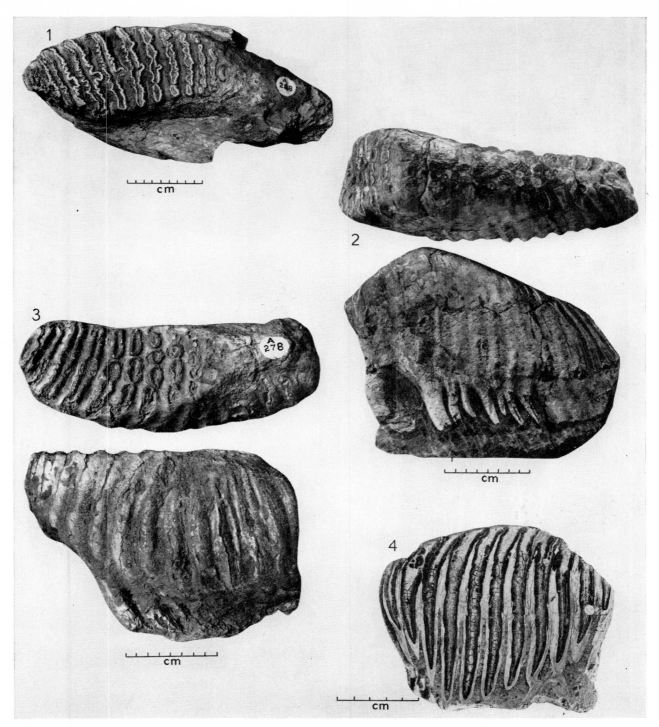

PLATE XIII

Elephas hysudricus Falconer and Cautley

1. GSI A288, left M_3. Siwalik Hills, India.
2. BM(NH) M-3129, right M^3. Siwalik Hills, India.
3. GSI A278, right M_3. Siwalik Hills, India.
4. BM(NH) M-3123, left M^1. Siwalik Hills, India.

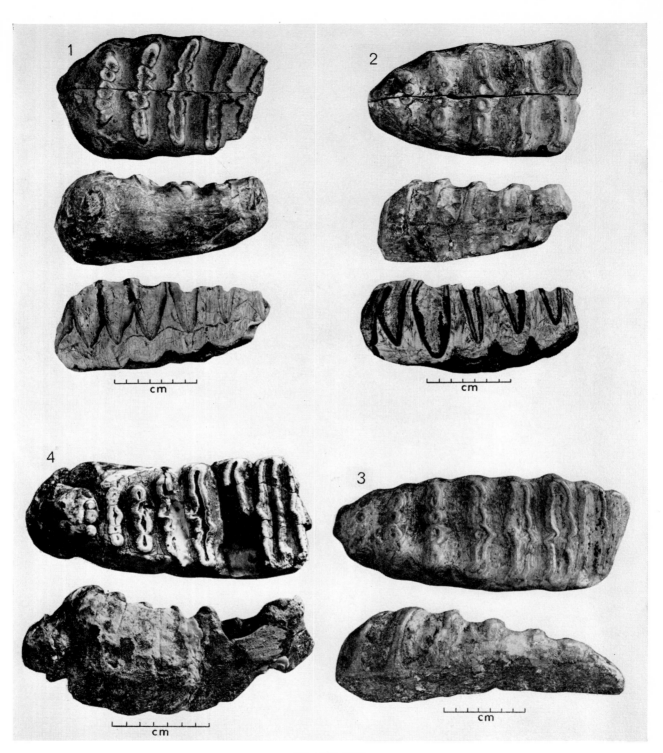

PLATE XIV

Mammuthus subplanifrons (Osborn)
1. MMK 3920, *Type*, right M_3. Vaal River, Sydney-on-Vaal, South Africa.
2. MMK 4334, right M_3. Vaal River, Gong-Gong, South Africa.
3. Same, left M_3.
4. NMB A2882, left M_3. Virginia, O.F.S., South Africa.

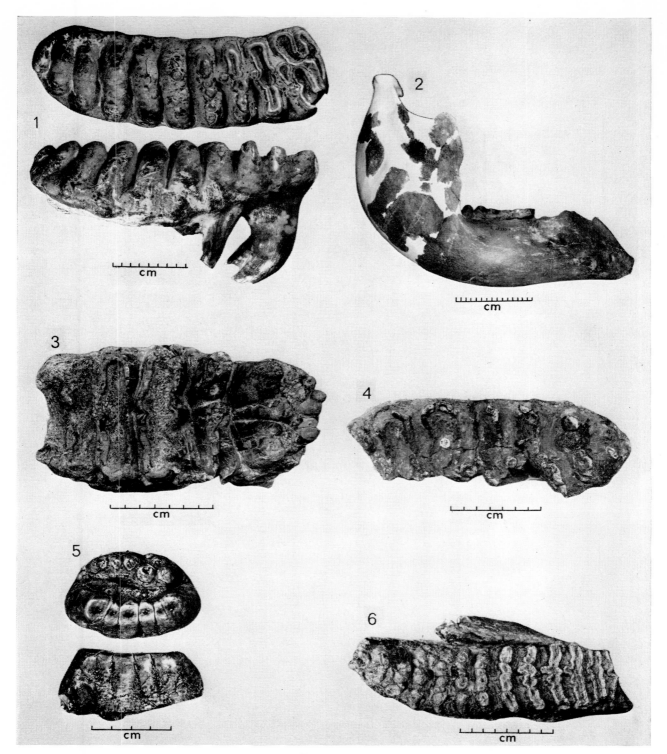

PLATE XV

Mammuthus subplanifrons (Osborn)

1. SAM L12723B, left M_3. Langebaanweg, Cape, South Africa.
2. SAM L12723C, right mandible. Langebaanweg, Cape, South Africa.
3. BM(NH) M-15410, left M^3. Kanam East, Kenya.
4. BM(NH) M-15411, left M_3. Kanam East, Kenya.
5. SAM 11714, molar fragment. Langebaanweg, Cape, South Africa.

Mammuthus africanavus (Arambourg)

6. MNHN (no number), left M_3. Koulá, Chad.

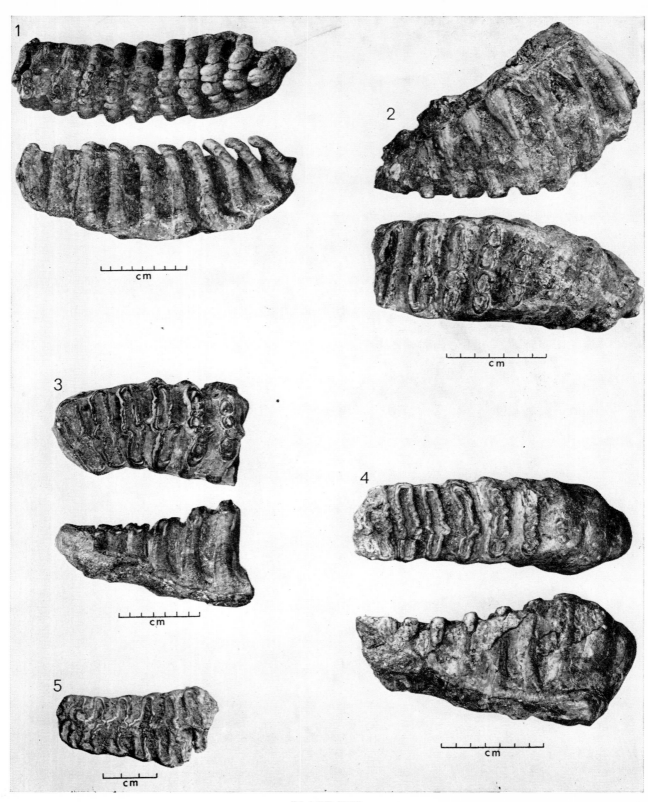

PLATE XVI

Mammuthus africanavus (Arambourg)

1. MNHN 1950-1-12, *Type,* right M_3. Ichkeul, Tunisia.
2. MNHN 1950-1-82, left M^2. Ichkeul, Tunisia.
3. MNHN 1950-1-98, left M_3. Ichkeul, Tunisia.
4. MNHN 1950-1-65, left M_2. Ichkeul, Tunisia.
5. MNHN 1950-1-76, right dm_4. Ichkeul, Tunisia.

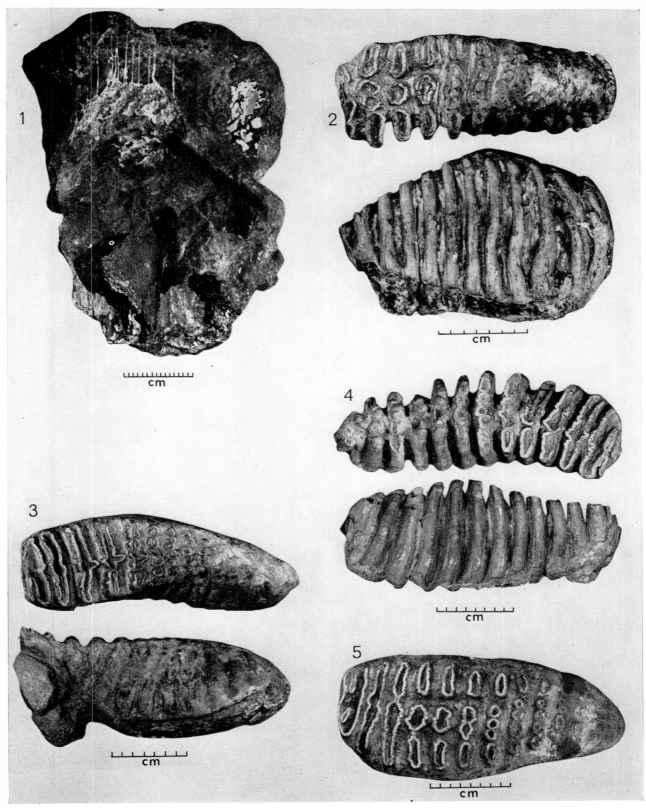

PLATE XVII

Mammuthus meridionalis (Nesti)

1. IGF 1051, skull. Upper Val d'Arno, Italy.
2. IGF 21, left M^2. Upper Val d'Arno, Italy.
3. IGF 33, left M_3. Upper Val d'Arno, Italy.
4. IGF 8, right M_3. Upper Val d'Arno, Italy.
5. MNHN 1892-15, left M^3. Senèze, France.

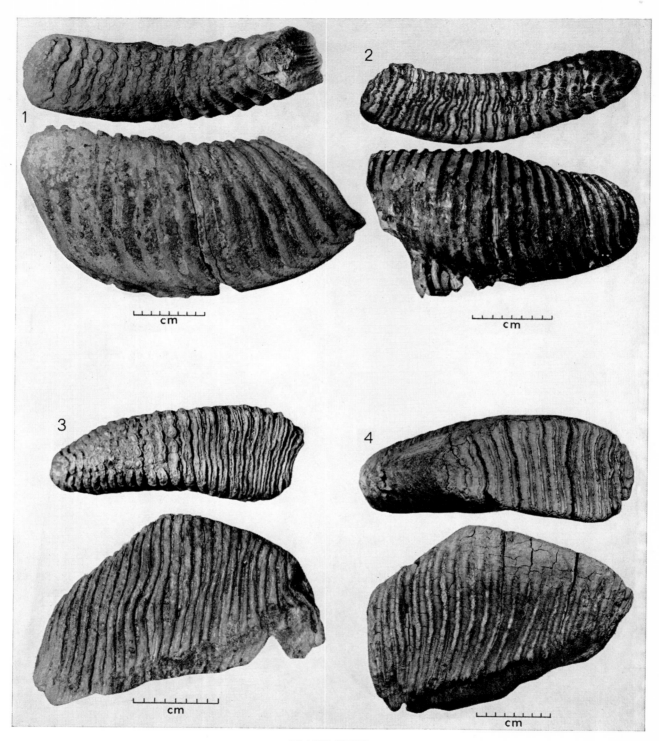

PLATE XVIII

Mammuthus armeniacus (Falconer)
1. NMB D.126a, right M_3. Süssenborn, Germany.

Mammuthus primigenius (Blumenbach)
2. UCB 66250, right M_3. Near Darmstadt, Germany.
3. NHB E.174, right M^3. Wasenboden, Switzerland.
4. NHB E.1, right M^3. Lörrach, West Germany.

XIII. INDEX

Bold-face type indicates a main entry; italics indicate that the reference is in a table or a figure.

Accessory columns, 16, 17, 19, 78, 79, **87–89**, 92, 93
Adaptive radiation, 111
adaurora, see *Loxodonta adaurora*
Aden Gulf, corridor across, 114, 118; faulting in, 112
Aegean corridor, 112
africana, see *Loxodonta africana*
africanavus, see *Mammuthus africanavus*
Afrochoerus nicoli, 72
Aguirre, E., 6, 11, 13, 16, 56, 59, 68, 78, 83
Ailuropoda, 75
Aïn Boucherit, 53, 68, *70*, 114
Aïn Brimba, *70*
Aïn Hanech, 24, 56, 57, *70*, 74, 87, 117
Alia Bay, 69
Allen, G. M., 29
Ambrosetti, P., 42
americanus (=*Mammuthus primigenius*), 60, 63
Anancinae, 5, 6, 113
Anancus, 16, 17, 69, 74; *osiris*, 69
andrewsi (=*Mammuthus subplanifrons*), 51
angammensis (=*Loxodonta africana*), 29, 31
Angular displacement of muscle fibers, 102, 103
Angus local fauna, 62, 63
antiquus (=*Elephas namadicus*), 31–33, 35, 37, 40–43, 53, 77, 79, 81–83, *93*; *antiquus recki* (=*Elephas recki*), 24, 35; *antiquus trogontheroides* (=*Mammuthus armeniacus*), 57, 59
Arambourg, C., 13, 24, 27, 29, 39, 53, 57, 68
Archidiskodon (=*Mammuthus*), 15, 23, 34, 37, 39, 42–45, 50, 51, 56, 62, 63, 64
archidiskodontoides (=*Elephas iolensis*), 37, 39
Arkalon local fauna, 62, *70*
armeniacus, see *Mammuthus armeniacus*
asiaticus (=*Elephas maximus*), 50
atlantica, see *Loxodonta atlantica*
atlantica atlantica, see *Loxodonta atlantica atlantica*
atlantica zulu, see *Loxodonta atlantica zulu*
ausonius (=*Elephas namadicus*), 40, 42
Australopithecine caves, 34, *70*, 73
Azzaroli, A., 56, 68, 73, 74

Baard's Quarry, 73
Bab el Mandeb Strait, opening of, 112
Bacton Forest beds, 56, *71*, 81, 87
Bahanga, see Kaiso Formation
Bakker, E. M. van Zinderen, 114
Balkans, Pleistocene climates in, 112
Baringo, Lake, 18, 20, 69
Barytheria, 15
Barytheriidae, 15
Barytherioidea, 5, 15
Barytherium, 15

Beausejour Farm, 37
Behrensmeyer, A. K., 72
Bel Hacel, *70*
Bering Strait, 118
Beru, 85
Bessarabia, 56
Bethlehem, 44, 45, *70*, 74, 84, 114
Betic-Riff corridor, 112, 117
Bishop, W. W., 20, 69, 72, 78, 79
Bizard, C., 31
Blancan Stage, 67, 76
Blumenbach, J. A., 29, 60, 61
Bogo, 49
Bolt's Farm, 34, *70*, 73, 114
bombifrons, see *Stegodon bombifrons*
Bos, 75; *primigenius*, 76
Boulder Conglomerate, *71*, 75, 114
Boule, M., and P. Teilhard de Chardin, 76
Brachypotherium, 68
brachyramphus (=*Mammuthus primigenius*), 60
Brain, C. K., 73
broomi (=*Elephas iolensis*), 37, 39
Brown, F. H., and K. R. Lajoie, 69
Brundon, 60
Bruneau local fauna, 62, 63, *70*, 118
Bugoma, see Kaiso Formation
Bumiaju, 46
Burg-Tonna, 60, 61
Burnett, G. T., 51
Butzer, K., 39

Camelus punjabicus, 75
Camper, P., 51
campylotes (=*Mammuthus primigenius*), 60
Canis, 74
capensis (=*Elephas africana*), 29
Carrière Sidi Abder Rahmane, 25, 29, 37, 39, *70*
Caspian region, Pleistocene vegetation in, 112
cautleyi, see *Stegolophodon cautleyi*
celebensis, see *Elephas celebensis*
Ceratotherium, 69; *simum*, 72
Cervus elaphus, 76; *ramosus*, 74
Chad, Lake, 53, 74, 114; Kolinga, 20, 22, *70*; Koulá, 20, 53, *70*; Ouadi Derdemi, 35, 53, *70*, 74, 81; Pleistocene climates of, 112; Yayo, 31, *70*
Chagny, 56, *71*, 74
Cheirolites (=*Mammuthus*), 50
Chelles, 57, *71*
Chemeron Formation, 20, 21, 23, 25, 35, 51, *70*, 72, 86; Locality J.M. 511, 20, 21, 51; Locality J.M. 514, 23; Locality J.M. 90, *70*, 72
Chemoigut beds, *70*
Chiwondo beds, 23, 34, 51, *70*, 74, 86, 114, 117
Choukoutien, 40, *71*, 75, 76
Clacton, 57, *71*

Colbert, E. H., 75; and D. A. Hooijer, 75
columbi, see *Mammuthus columbi*
commutatus (=*Mammuthus primigenius*), 60
Cooke, H. B. S., 11–13, 25, 27, 34, 39, 68, 72, 73; and S. Coryndon, 20, 25, 37, 72; and R. F. Ewer, 69; and V. J. Maglio, 6, 72
Coppens, Y., 17, 20, 23–25, 31, 37, 53
Coralline Crag, *71*
Cornielia, *70*
Correlation of Pliocene/Pleistocene deposits, Africa, 68–73; Asia, *71*, 74–76; Europe, *71*, 73–74; North America, *70*, 76
Corton Sands, 57, *71*
Coryndon, S., 72
Cox, A., 76
Crete Formation, 63, *70*
Crocuta, 75
Cromer Forest beds, 42, 57, 59, *71*, 75, 117
Cromerian, 73, 83, 106
Crooked Creek Formation, *70*
Crusafont-Pairó, M., and S. Reguant, 14
Curtis, G. H., 75
Cuvier, G., 29, 60
cyclotis (=*Loxodonta africana*), 29
cypriotes, see *Elephas cypriotes*

darti (=*Elephas iolensis*), 37
darwins, *105, 106, 107, 108*
Deep masseter muscle, function of, 103; position of, *101*
de Heinzelin, J., 114
Deinotheriidae, 15
Deinotherioidea, 15
Deinotherium, 15
Depéret, C., and L. Mayet, 55, 59, 60
Deraniyagala, P.E.P., 15, 33
Desa Beru, 46
Dhok Pathan, *71*, 75
Dicerorhinus etruscus, 73
Dicyclotherium (=*Mammuthus*), 60
Dien, M. N., and L. P. Chia, 75
Dietrich, W. O., 37, 56, 69
Digastric muscle, function of, 103; position of, *101*
Djambe, 49
Djetis beds, 44, 46, *71*, 75, 85, 114
Durfort, 56, *71*
Dwarfing in elephants, 46, **94**, 106, 107

eellsi (=*Mammuthus columbi*), 63
Ehringsdorf, 57, 60, *71*, 76
Ekora, 17, 20, 23, 33, 34, *70*, 74, 81, 84
ekorensis, see *Elephas ekorensis*
Elandsfontein, 25, 27, 29, *70*, 73, 81, 114
Elephantidae, 5, 15, **16**, 46, 89
Elephantinae, 15, 16, **20**
Elephantoidea, 15, 16
Elephantus (=*Elephas*), 31
Elephas, *15*, 16, 20, **31–33**, 34, 35, 37, 40, 42, 44–47, 49–53, 64, 65, 67, 75, 81, *82*,

145

83–86, 94–97, 99, 100, *105*, 106, 108, 111, 114, *115*, 117–119; *celebensis*, 32, **46**, *47*, 65–67, *77, 79, 84*, 85, *93*, 94, *115*, 118; *creticus*, 42; *cypriotes*, 42; *ekorensis*, 32, **33–34**, 35, 36–38, *44*, 46, 64–65, 67, 69, 74, 77, 79, 81, *82*, 83–85, 87, 93, 95, 96, 106, *108*, 114, *115*, 118; *falconeri*, 31–33, **42**, 66, *77, 79, 82, 83, 93*, 107, *115*, 118; *hysudricus*, 32, 33, 42, 47, **48–49**, 50, 65–67, 75, *77, 79*, 83–86, 93, *95–98*, 100, 107, 114, *115*, 118; *hysudrindicus*, 32, **49–50**, 66, 67, *77, 79, 84*, 85, *115*, 118; *iolensis*, 29, 32, **37–40**, 66, 67, 73, *77, 79, 81, 82, 93*, 94, 95, *96*, 106, 108, 114, 115, *117*, 118; *lamarmorae*, 42; *maximus*, 31, 32, 49, **50**, 66, 67, *77, 79, 83, 84*, 85, 86, *90, 93*, 94, *95–98*, 100, *104*, 107, 115, 118; *melitensis*, 42; *mnaidriensis*, 42, 83; *namadicus*, 31, 33, 37, 39, **40–42**, *43*, 49, 59, 66, 67, 75, *77, 79*, 81–83, *94, 96*, 100, 106, 115, 117, 118; *planifrons*, 31–34, **42–46**, 47–49, 51, 53, 55, 56, 65–67, *74*, 75, *77, 79*, 83–85, *90, 93, 94, 96*, 100, 114, 115, 117, 118; *platycephalus*, 31, 33, **47**, *77, 79, 84*, 118; *recki*, 29, 32, 33, **34–37**, 38, 39, 41, 49, 65, 67, *72, 74, 77, 79*, 81–83, 85, *93*, 95, *96*, 100, 106, 108, 114, *115*, 118
Elephasidae (=Elephantidae), 16
Emmendingen, 60
Enamel, folding of, 10, 91; leading edge of, 10, *90, 91, 93*; thickness of, 8, **10**, 13, 89, 92, 94, *106*
Equus, 45, 67, 69, 72–75, 87; *hemionis*, 76; *przewalskii*, 76; *sanmeniensis*, 75; *sivalensis*, 75; *yunanensis*, 75
Ericson, D. B., and G. Wollin, 76
Erpfingen, *71*, 74
Etouairés, *71, 74*, 75
Evernden, J. F., *et al.*, 118
Evolutionary rate, 105–111; in *Elephas*, *108*; in enamel thickness, *106*; in hypsodonty index, *106*; in *Mammuthus*, *107*; in plate number, *105*
Evolutionary trends, in the dentition, 87–95; in the mandible, 95, 97; in the masticatory apparatus, 101–105; in the skull, 97–101
Ewer, R. F., 73
exilis (=*Mammuthus columbi*), 63
exoptatus (=*Elephas recki*), 23, 34, 35
Extinction, rate of, *109, 110*, 111

Falconer, H., 29, 31, 41, 59, 60, 62
Falconer, H., and P. T. Cautley, 41, 42, 43
falconeri, see *Elephas falconeri*
Farnetta, 56, *71*
Fauna Antiqua Sivalensis, 41
Faunal spans, 68
Fejfar, O., 74
Felis, 74; *tigris*, 75; *pardus*, 76
Ferladani, 56
Fernex, F. C., *et al.*, 112
First appearances, rates of, 108, *109, 110*
Fitch, F. and J. A. Miller, 72
Flerov, K. K., 74
Flint, R. F., 73
floridanus (=*Mammuthus columbi*), 63
Fouarat, 53, 68, *70*, 117

Franz, H., 112
Frenzel, B., 112, 117
Functional length, molar, 11

Galana Boi beds, *70*
ganesa, see *Stegodon ganesa*
Garet et Tir, 24, 53
Gaudry, A., 6, 7, 18
Gazella, 69
Gazellospira torticornis, 73
Genus concept, 14
Gibraltar Strait, 12; migration corridors across, 117, 119
Giermann, G., 112
giganteus (=*Mammuthus primigenius*), 60
gigas (=*Elephas maximus*), 50
Gignoux, M., 112
Godavari Valley, 40, *71*, 117
Goldfuss, G. A., 29, 31
Gomphotheriidae, 5, 7, 15, 16, 46, 113; dentition of, 79, **87–89**, 92; skull of, *18*, 78, 98, 99
Gomphotherioidea, 16
Gomphotherium, 69, 78, 87, 96, 98, 99; *angustidens*, 18, *78, 88–90, 93*
gomphotheroides, see *Primelephas gomphotheroides*
Gray's Thurrock, 29, 31, 40
Gregory, W. K., 97
Grenada, 56
griqua (=*Elephas recki*), 34, 64
Griqualand West, 64
Guenther, E. W., 74

Haas, G., 83
Hammon, A. H., 112
hanekomi (=*Elephas iolensis*), 37
haroldcooki (=*Mammuthus meridionalis*), 62, 63
Harris, J., 15
hayi (=*Mammuthus meridionalis*), 62, 63
Height of molar, 8, 10, 11, **12**, 91, 92, 94
Heintz, E., 73
Hemibos triquetricornis, 75
Hendey, Q. B., 72, 73
Hesperoloxodon (=*Elephas*), 15, 31, 33
Hibbard, C. *et al.*, 76
Hill, A., 72
Hipparion, 45, 72, 74, 75
Hippopotamus, 68; *immagunculus*, 69; *protamphibius*, 72
Hippotragus, 72
Holloman's Quarry, 62, 63
Homa Mountain, 35, 37, *70*
Homo sapiens, 76
Hooijer, D. A., 24, 42, 44, 45, 46, 53, 72, 74, 75, 85; and B. Patterson, 68, 69
Hopwood, A. T., 37
Horizontal shearing, 9, **89–91**
Hoshangtung Cave, *71*, 75
Houghton, S. H., 64
Howell, F. C., 68, 73
Hsiachaohwan, *71*, 114
Hürzeler, J., 73
Hyaena, 69, 74, 75
Hylobates, 75
Hypselephas (=*Elephas*), 15, 32, 33

hypsilophus (=*Elephas celebensis*), 46, 47
Hypsodonty index, 9, **10–11**, 13, *106*
Hystrix, 74
hysudricus, see *Elephas hysudricus*
hysudrindicus, see *Elephas hysudrindicus*

Ichkeul, 24, 53, 68, *70*, 74, 87, 114, 117
Ileret, 37, *70*, 72, 74, 81
imperator, see *Mammuthus imperator*
Incisors, mandibular, 16–20, 23, *32*, 46, 96
Index fossils, use of, 67
indicus (=*Elephas maximus*), 47, 49
indonesicus (=*Elephas celebensis*), 32, 46
insignis, see *Stegodon insignis*
intermedius (=*Mammuthus armeniacus*), 57, 59
Intravalley columns, see Accessory columns
iolensis, see *Elephas iolensis*
Irrawaddy Beds, Lower, *71*, 75; Upper, 49, *71*, 75, 85, 114, 115
Irvingtonian Stage, 62; faunas of, 118

jacksoni (=*Mammuthus primigenius*), 63
Janossy, D., 74
Jaramillo normal polarity event, 76
jeffersonii (=*Mammuthus columbi*), 62, 63
Jepsen, G. L., 14
Jisr Banat Yaqub, 42
jubatus (=*Mammuthus primigenius*), 60

Kahlke, H. D., 74
Kaiso Formation, 35, 69, *70*, 74, 81, 117; Bahanga, 37; Bugoma, 23, 25; Kaiso Village, 37, 72; Nyabrogo, 20, 21, 51, 72; Nyawiega, 20, 21, 72, 80
Kaiso Village, see Kaiso Formation
Kali Glagah, *71*, 75
kamensis (=*Mammuthus primigenius*), 60
kamenskii (=*Mammuthus primigenius*), 60
Kanam, 20, 21, 22, 51, 52, *70*, 86, 117
Kanapoi, 3, 17, 23, 25, 34, 69, *70*, 72–74, 80, 81
Kanjera, 35, *70*, 81
Kaperyon Beds, 18, 20, 69, *70*, 79
Karewa Beds, 42, 49, *71*, 85, 117
Kebili, 53, 68, *70*, 114
Kedoeng, 115
Kedoeng Broeboes, 49
Kedoeng Loemboe, 49
Kent, P. E., 69
Key to species of Elephantidae, 66–67
Kibish Formation, 37, *70*
Kikagati, 35, *70*, 81
Kobus, 72
Kolinga, see Chad
Koobi Fora, 23, 25, 35, 37, *70*, 72–74, 80, 81
korotorensis, see *Primelephas korotorensis*
Koulá, see Chad
Kretzoi, M., 46, 74
Kromdraai, *70*, 73
Kubi Algi, 23, 25, 34, 69, *70*, 81, 84, 85
kuhni (=*Elephas iolensis*), 37, 39

Kummel, B., 112
Kurtén, B., 73, 74

Laetolil, 24, 35, 37, 69, *70,* 72
Laiatico, 56, 57, 65, *71,* 74, 87, 117
Lamellae number, see Plate number
Lamellar frequency, **9**, 10, **12**, 13, 93
Langebaanweg, 21, 51–53, *70,* **72**, 73, 86, 87, 117
l'Aquilla, 56
Last appearances, 108, *109,* 110
Latemné, 42, *70,* 115
Lateral pterygoid muscle, position of, *99,* 100, 101; function of, *102,* 103, 104
Leakey, L. S. B., 69, 72
Leakey, R. E. F., 34, *72*
Leffe, 71
Lehman, U., 74
Leidy, J., 62
Leith-Adams, A., 67, 60
Leith-Adamsia (=*Elephas*), 31, 33
Length-Lamellae ratio, 12
Length of molar, 8, *11,* 89, 91
Length/width index, 9, **11**
Leptobos, 75; *etruscus,* 73; *falconeri,* 67, 75; *stenometopon,* 73
Lothagam Hill, 3, 16, 19, 20, 21, 22, 23, 24, 68, 69, *70,* 72, 80
Loxodonta, 6, 15, 16, 20, **22–23**, 24, 25, 29, 31–34, 37, 46, 51, 53, 64–66, 69, *80,* 81–83, 86, 87, 93–95, 97, 98, 100, 105–107, *113,* 114, 117–118; *adaurora,* **23–25**, 26, 29, *31, 32,* 44, 46, 53, 63–65, 67, 69, 72, *77, 79,* 80, 81, 86, 87, 92–100, *105, 113,* 114, 118; *africana,* **29–31**, 32, 65, 66, *77, 79, 80,* 81, 93–95, 97–101, 103, 105, 118; *atlantica,* 22, 23, **25–27**, *28,* 29–31, 39, 65–67, *77, 78, 80,* 81, 83, 93–95, 105, 106, *113,* 114, 118; *atlantica atlantica,* **27**; *atlantica zulu,* **25–27**
loxodontoides, Archidiskodon, 64
Ludolf, H. W., 60
Lutra, 74, 75
lybicus (=*Stegotetrabelodon syrticus*), 17, 18
Lydekker, R., 29, 43, 48
lyrodon (=*Mammuthus meridionalis*), 55

Ma Kai Valley, *71,* 75
Maarel, F. H. van der, 46
Macaca, 75
MacInnes, D. G., 23, 24
macrorhynchus (=*Mammuthus primigenius*), 60
Madoqua, 72
Maglio, V. J., 7, 14, 16, 17, 20, 25, 34, 35, 44, 46, 52, 56, 68, 69, 72, 78, 83, 89; and Q. B. Hendey, 5, 7, 16, 21, 53, 73
Maison Carrée Formation, 29
Makapansgat, 34, *70,* 73
Makiyama, J., 42
Malafrosca, 40
Malde, H. E., and H. A. Powers, 62, 118
Mammonteus (=*Mammuthus*), 51
mammonteus (=*Mammuthus primigenius*), 60
Mammontinae, 51
Mammontovoi Kost, 60
mammouth (=*Mammuthus primigenius*), 60
Mammut, 98, 99; *americanus,* 99
Mammuthus, 5, 10, 15, 16, 20, 24, 32, 33, 37, **50–51**, 53, 56, 57, 60, 61, 64, 65, 67, 74, 86, 87, 94–97, 100, 105, 106, 111, *116,* 117–119; *africanavus,* 35, 44, 50, 51, **54–55**, 56, 63, 65, 67, 74, *77, 79, 86, 89, 93,* 95, 96, *107, 116,* 117, 118; *armeniacus,* 50, 51, 56, **57–60**, 62, 66, 67, *77, 79,* 81, 86, 87, *93,* 95, 96, 107, *116,* 118; *columbi,* 50, 51, 62, **63**, 66, 67, *77, 79, 86,* 87, *93,* 116, 118; *imperator,* 50, 51, 62, **63**, 66, 67, *77, 79, 86,* 87, *116,* 118; *meridionalis,* 34, 40, 41, 44, 45, 50, 51, **53–57**, 58–60, 62, 63, 65–67, 73, 74, *77, 79,* 81, *86, 87, 93, 95–98,* 106, *107, 116,* 118, Laiatico Stage, **56**, *57, 77,* 117, Montavarchi Stage, **56**, 57, 73, 74, *77,* 117, Bacton Stage, **56**, 57, 62, *77;* *primigenius,* 50, 51, 58, **60–63**, 66, 67, *77, 79, 86,* 87, *89, 92,* 95, *96, 97, 98, 100, 107, 116,* 117, 118; *subplanifrons,* 50, **51–53**, 64, 65, 67, 73, *77, 79, 86,* 87, *89, 93, 95–97,* 106, *107, 116,* 116, 117, 118
Mammutidae, 5, 6, 15
Mammutoidea, 15, 16
Mandible, evolution of, 97, *99;* structure of, 95–96, *97*
Martin, P. S., and W. E. Wright, 111
Mastodontidae, 15
Mastodontoidea, 15
Matschie, P., 29
Matsumoto, H., 33, 42
Mattauer, M., 112
Mauer, 40, 42, 57, *71,* 117
Mawby, J., 34
maximus, see *Elephas maximus*
maximus fossilis (=*Elephas hysudrindicus*), 49
Mayr, E., 13, 14
Medial pterygoid muscle, function of, *101, 102;* position of, 101, 103
Median cleft, 87, *88,* 89, 92, *93*
Mediterranean basin, corridors across, *117;* geological history of, 111–112; dwarf elephants of, 42, 117
Megaloceros, 74
Meganthropus, 75
Meiring, A. J. D., 53
melitensis, see *Elephas melitensis*
Mellivora, 75
meridionalis, see *Mammuthus meridionalis*
meridionalis nebrascensis (=*Mammuthus meridionalis*), 62, 63
meridionalis voigtstedtensis (=*Mammuthus meridionalis*), 56
Mesochoerus, 69, 72; *limnetes,* 72; *olduvaiensis,* 72
Metarchidiskodon, 15, 64
Milk dentition, 89
milletti, Archidiskodon, 63
Miocene/Pliocene boundary, 3
mnaidriensis, see *Elephas mnaidriensis*
Modjokerto, 75
Moeritheriidae, 15
Moeritherioidea, 5, 15
Moeritherium, 15
moghrebiensis (=*Mammuthus meridionalis*), 57

Mogok Caves, 40, 42, *71,* 75
Montavarchi group, 56, 63
Montopoli, 56, *71,* 117
Morphological characters, definition of, 8–11; measurement of, 11–13
Mosbach, 40, 42, 57, 59, *71,* 74, 75, 117
Mount Coupet, 68, *71,* 74
Mpesida beds, 18, 20, 69, *70*
Mursi Formation, 23, 24, 25, 69, *70,* 73, 80
Musil, R., 74
Mwenirondo I, 34, *70*
Mwimbi North, 34, *70*

namadicus, see *Elephas namadicus*
namadicus naumanni (=*Elephas namadicus*), 31
Narbadda Valley, 40, 42, *71,* 75, 76, 117
Natodomeri I and II, 37
nestii (=*Mammuthus armeniacus*), 57
Ngandong, 49, *71,* 75, 85, 117
Ngorora, *70,* 78
Nihowan, *71,* 75
Niobrara River, 62, 63
Notochoerus capensis, 69
Nuchal musculature, 105
Nyabrogo, see Kaiso Formation
Nyanzachoerus, 68, 69
Nyawiega, see Kaiso Formation

odontotyrannus (=*Mammuthus primigenius*), 60
Olduvai Gorge, 35, 37, 69, *70,* 72–74, 81
Olduvai normal polarity event, 76
Olorgesailie, 35, 37, *70,* 81
Omo Valley (also see Shungura Formation), 69, *70,* 72, 73, 81
Omoloxodon (=*Elephas*), 32, 33
orbus, see *Stegotetrabelodon orbus*
Orontes Valley, 42
Osborn, H. F., 6, 15, 22, 25, 29, 33, 39, 40–44, 47, 51, 55, 60, 61, 62, 64
Ouadi Derdemi see Chad
Oued Akrech, 53
Oued Constantine, 25, 27, 39, *70*
oxyotis (=*Loxodonta africana*), 29

Palaeoloxodon (=*Elephas*), 15, 25, 31, 33
paniscus (=*Mammuthus primigenius*), 60
Paracolobus baringensis, 72
Pardines, *71,* 74, 75
Parelephas (=*Mammuthus*), 15, 50, 51
Parmularius, 72
Pastonian Crag, 59
Patterson, B., 3, 69
Pavlow, M., 56
Pei, W. C., 75
Peninj, *70*
periboletes (=*Mammuthus primigenius*), 60
Petrocchi, C., 16, 17, 18, 68, 78
Phacochoerus, 73
Phyletic branching, 108, *109,* 111
Phyletic evolution, *109,* 111
Phylogeny, of Elephantidae, present concept, 76–87; previous views, 6, 7; of proboscidean families, 6

Pian dell'Ollmo, 40, 41
Piccard, L., 114
Pignataro Interamna, 40, 41, 42
Pilgrim, G. E., 42, 75
Pilgrimia (=*Elephas*), 31, 33, 34, 39
Pinjor, 44, 45, 49, *71*, 75, 85, 115
planifrons, see *Elephas planifrons*
planifrons nyanzae (=*Mammuthus subplanifrons*), 51
planifrons rumanus (=*Mammuthus meridionalis*), 55
Plate number, **9**, **11**, *105*
Plate spacing, see Lamellar frequency
Platelephas (=*Elephas*), 15, 31, 33, 47
platycephalus, see *Elephas platycephalus*
platyrhynchus (=*Elephas namadicus*), 40, 42
platytaphrus (=*Mammuthus primigenius*), 60
Pliocene/Pleistocene, boundary, 3; correlation of deposits, 67–76
Pohlig, H., 33, 59
Polyphyletic origin of Elephantidae, 46
Pomel, A., 25, 38, 39
pomeli (=*Loxodonta atlantica atlantica*), 25, 29, 37, 39
Pongo, 75
Port de Mostaganem, 37, 39
Posttrite cones, *88, 89*
praeplanifrons (=*Elephas planifrons*), 42, 43, 44
Praetiglian, 56, *71*, 74
Předmosti, *71*, 74, 76
Premolars, in elephants, 89
Pretrite cones, *88, 89*
prima (=*Loxodonta africana*), 29
primaevus (=*Mammuthus primigenius*), 60
Primelephas, 15, 16, **20**, 52, 64, 66, 69, 72, 80, 86, 87, 113, 118; *gomphotheroides*, 20, **21–22**, 46, 64, 66, *77*, *79*, 80, 82, 84, 86, 89, 90, 92–96, 105, 106, 107, 108, 118; *korotorensis*, 20, 21, **22**, 64, 66, 70, *77*, 118
primigenius, see *Mammuthus primigenius*
priscus (=*Loxodonta africana*), 29, 31, 40
progressus (=*Mammuthus columbi*), 63
proplanifrons (=*Mammuthus subplanifrons*), 51
protomammonteus (=*Elephas namadicus*), 40, 42
pygmaeus (=*Mammuthus primigenius*), 60

Quezel, P. and C. Martin, 114

Rabat Casablanca beds, 69, *70*
Ramaccioni, G., 56
Rancholabrea, *70*, 118
Rates of evolution, 105–111
Rawe, *70*
recki, see *Elephas recki*
Red Crag, *71*, 117
Rhinoceros, 75
Rocca Neyra, 68, *71*, 75
roosevelti (=*Mammuthus columbi*), 63
Ruddiman, W. F., 76

Sablière de Palekao, 25
Sahabi, 16, 18, 19, 68, *70*, 78, 80
sahabianus (=*Stegotetrabelodon syrticus*), 17, 18
Saint Valliers, *71*, 74
Sainzelles, 68, *71*, 74
Sangiran, 49, 115
San Paolo de Villafranca, 59
San Romano, 40
Savage, D. E., and G. H. Curtis, 74, 76
Savage, R. J. G., 111
Scott, W. B., 25, 27
scotti (=*Mammuthus subplanifrons*), 51
Senèze, 56, *71*, 73
Shansi, 44, 114
Shear angles, movement of, 90
Shearing index, definition of, *10*, **93**; evolution of, 94–95
sheppardi (=*Elephas iolensis*), 37, 39
Shungura Formation, 25, 29, 31, 34, 35, 37, 69, *70*, 72–74, 80, 81, 114
sibericum (=*Mammuthus primigenius*), 60
Sidi Abder Rahmane, see Carrière Sidi Abder Rahmane
Simpson, G. G., 13, 15, 16, 51, 76
Singer, R., and D. A. Hooijer, 53
Sinomalayan fauna, 75
Sivacobus palaeindicus, 75
Sivalikia (=*Elephas*), 33
Sivamalayan fauna, 74, 75
Sivatherium giganteum, 75
sivalikiensis (=*Elephas planifrons*), 31, 33, 42, 43
Siwalik Hills, 24, 33, 44, 45, 49, 53, 75, 114
Sjara-osso-Gol, 40, *71*, 75, 76, 117
Skull, evolution of elephant, 99–101; structure of elephant, 97–99
Soergel, W. E., 60, 67
Solo Valley, 49
Sompoh, 46, 85
sonorensis (=*Mammuthus imperator*), 63
Species concept, 13, 14
Species duration, 110, 111
Species of elephant, time distribution, *79*
Species turnover rate, *110*, 111
Spinegallo, 42, *71*
Stamp, L. D., 75
Stegodibelodon schneideri, 17
Stegodon, 5, 6, 15, 16, 20, 47, 75, 78; *bombifrons*, 98; *ganesa*, 98; *insignis*, 75, 78, 98
Stegolophodon, 6, 7, 15, 16, 17, 51, 53, 73, 78; *cautleyi*, 78
Stegoloxodon (=*Elephas*), 32, 33
Stegotetrabelodon, 15, **16–17**, 22, 64, 69, *77*, 78–80, 96, 113, 118; *orbus*, 16, 17, **18–20**, 32, 64, 66, *77*, *79*, 80, *90*, *92*, *93*, 118; *syrticus*, 16, **17–18**, 19, 64, 66, *77*, 78, *79*, 80, 89, *93*, 96, 118
Stegotetrabelodontinae, 15, **16**, 80, 112
Steinheim, 40, 57, *71*
stenotaechus (=*Mammuthus primigenius*), 60
Sterkfontein, 34, *70*, 73, 114
Stross, F. H., 75
Stylohipparion, 69, 72
subantiqua (=*Elephas iolensis*), 37

subplanifrons, see *Mammuthus subplanifrons*
Successional species, *109*, 110, 111
sumatranus (=*Elephas maximus*), 50
Superficial masseter muscle, function of, 103–105; position of, 101
Sus, 74, 75; *hysudricus*, 75
Süssenborn, 42, 57, *71*, 74, 117
Swanscombe, 57, *71*
Swartkrans, *70*, 73
Swartz, D. H., and D. D. Arden, 112
Sydney-on-Vaal, 51
Symmetry in elephant molars, evolution of, 27–28
syrticus, see *Stegotetrabelodon syrticus*
Szechuan, see Yenchingkou

Tapirus augustus, 75
Tatrot horizon, 44, 45, *71*, 75, 84, 85, 114
Taubach, 40, 42, 57, *71*, 75, 76
Taung, *70*
Tchernov, E., 114
Tegelen Clays, *71*, 74
Teilhard de Chardin, P., 75
Temporalis muscle, function of, 101, 103, 104; position of, 98, 99, 100, *102*
Ternifine, 25, 27, 29, 69, *70*, 81, 114
Thenius, E., 74
Thomas, G., 13
Tinggang, 49, 85
Tjidoelang, 44, 46, *71*, 75, 114
Tobias, P. V., 83
Tokunaga, S., 42
tokunagai (=*Elephas namadicus*), 39, 40, 42
Toungour, 35, 53
transvaalensis (=*Elephas iolensis*), 37, 39
Trinil, *71*, 75, 117
trogontherii (=*Mammuthus armeniacus*), 42, 57, 59
Trogontherium cuvieri, 76
Turnover rate in elephant species, 108, *110*, 111

Ubeidiya, 83, 115
Upnor, 40
Ursus, 75; *arctos*, 76; *etruscus*, 73, 75; *malayanus*, 75; *spelaeus*, 76
Usno Formation, *70*, 74

Vaal River, 37, 51, 63, 64, *70*, 73, 86
Val d'Arno, 56, 63, *71*, 73, 74
Van Bemmelen, R. W., 73
Van Valen, L., 111
vanalpheni, *Archidiskodon*, 63
Velay, 74
Viallette, *71*, 74
Villafrancha d'Asti, 45, *71*, 74
Villafranchian Stage, 3, 73, 74, 75, 76, 83; definition of, 68; in Africa, 87; in Europe, 45, 81, 117; index fossils of, 44, 56, 57, 67
Villaroya, 45, *71*, 73, 74
Virginia, O.F.S., 51, 53
Viverra, 75
Vogel River Series, 23, 24, 25, 35, 37, 51, *70*, 72
Voigtstedt, 56, *71*, 74

von Koenigswald, G. H. R., 44, 75
Vondra, C. F. et al., 34
Vulpes, 74

Wadi Natrun, *70*
washingtonii (=*Mammuthus columbi*), 63
Weithofer, K. A., 67, 97
Weybourne Crag, 68, *71*
Width, molar, 11, **12**

wilmani (=*Elephas iolensis*), 37, 39
wüstii (=*Mammuthus armeniacus*), 57

Yayo, see Chad
Yellow Sands, see Mursi Formation
Yenchingkou, 40, 42, *71*, 75, 117
yokohamanus (=*Elephas namadicus*), 40, 42
yorki, *Archidiskodon* (=Elephantinae, gen. indet.), 37

yorki, *Pilgrimia* (=*Elephas iolensis*), 37, 39
Yüshe, *71*, 75

Zeuner, F. E., 73, 74, 112
Zoogeography, of *Elephas*, 114, *115*, 117; of *Loxodonta*, *113*, 114; of *Mammuthus*, *116*, 117, 118
Zuffardi, P., 58, 59
zulu, see *Loxodonta atlantica zulu*